Retroviral Testing
Essentials for Quality Control and Laboratory Diagnosis

Niel T. Constantine, Ph.D., M.T. (ASCP)
Associate Professor and Director
Clinical Immunology Laboratory
Department of Pathology
University of Maryland Medical System/Hospital
Baltimore, Maryland

Johnny D. Callahan, B.S., M.T. (ASCP)
Research Assistant
University of Maryland
Baltimore, Maryland

Douglas M. Watts, Ph.D., M.S.
Infectious Diseases Threat Assessment Program
U.S. Naval Medical Research Institute
Bethesda, Maryland

CRC Press
Boca Raton   Ann Arbor   London   Tokyo

Library of Congress Cataloging-in-Publication Data

Constantine, Niel T.
    Retroviral testing: essentials for quality control & laboratory diagnosis / authors, Niel T. Constantine, Johnny D. Callahan, Douglas M. Watts.
      p. cm.
    Earlier version published with title: HIV testing & quality control. 1991.
    Includes bibliographical references and index.
    ISBN 0-8493-4429-8
    1. AIDS (Disease)--Serodiagnosis--Quality control. 2. HIV (Viruses)--Analysis--Quality Control. 3. Diagnosis, Laboratory--Quality Control. I. Callahan, Johnny D. II. Watts, Douglas M. III. Constantine, Niel T. HIV testing & quality control. IV. Title.
    [DNLM: 1. AIDS Serodiagnosis--methods. 2. HIV Infections--diagnosis. 3. Quality Control. 4. Retrovirus Infections--diagnosis. WD 308 C758r]
    RC607.A26C675 1992
    362. 1'969792--dc20
    DNLM/DLC
    for Library of Congress

                                                              92-15536
                                                                    CIP

    The views expressed in this book are those of the authors and not necessarily those of the organizations with which they are affiliated.

    Support for the publication of an earlier version of this book (*HIV Testing and Quality Control: A Guide for Laboratory Personnel,* ISBN 0-939704-07-2) was provided by AIDSTECH/Family Health International with funds from the United States Agency for International Development (USAID). Additional support for the earlier version was provided by the United States Naval Medical Research and Development Command, Bethesda, MD, Work Unit No. 3M463105.H29.AA335.

    This book represents information obtained from authentic and highly regarded sources. Reprinted material is quoted with permission, and sources are indicated. A wide variety of references are listed. Every reasonable effort has been made to give reliable data and information, but the author and the publisher cannot assume responsibility for the validity of all materials or for the consequences of their use.

    Neither this book nor any part may be reproduced or transmitted in any form or by any means, electronic or mechanical, including photocopying, microfilming, and recording, or by any information storage and retrieval system, without permission in writing from the publisher.

    Direct all inquiries to CRC Press, Inc., 2000 Corporate Blvd., N. W., Boca Raton, Florida, 33431.

                              © 1992 by CRC Press, Inc.

                  International Standard Book Number 0-8493-4429-8

                        Library of Congress Card Number 92-15536
                  Printed in the United States 1 2 3 4 5 6 7 8 9 0
                            Printed on acid-free paper

AUTHORS

Niel T. Constantine, Ph.D., M.T. (ASCP), is an Associate Professor at the University of Maryland School of Medicine, and Director and Division Head of Clinical Immunology at the University of Maryland Medical System/Hospital in Baltimore, Maryland. Dr. Constantine obtained his Ph.D. at the University of Maryland School of Medicine Department of Pathology in 1981.

Dr. Constantine has taught medical technology students for over eight years, and was course chairman for clinical immunology, anatomy and physiology, and clinical pathology at the Program in Medical Technology at the University of Maryland. He also acted as assistant director for continuing education.

In 1987, Dr. Constantine accepted an overseas assignment at the U.S. Naval Medical Research Unit #3 in Cairo, Egypt to direct a diagnostic and research laboratory for retro- and hepatotropic viruses. During his three years in Cairo, the laboratory became a World Health Organization (WHO) collaborating center for AIDS in the Eastern Mediterranean region. Dr. Constantine coordinated many HIV testing workshops and acted as a WHO consultant in several East African and Middle East countries. When returning to the University of Maryland in 1990, his major responsibilities include directing the clinical immunology laboratory in a 800 bed hospital, managing the budget, and conducting research in the field of retrovirology. His major research interests include the development of improved diagnostic assays for the retroviruses.

Dr. Constantine is a member of the American Society of Clinical Pathologists, and is active in national and international conferences. He is the recipient of several research grants from the World AIDS Foundation to conduct HIV technology transfer in the countries of Ethiopia and Rwanda. His past activities included consultancies in Kuwait, Egypt, Pakistan, Abu Dabi, and Ethiopia. Recently, he has acted as a consultant for Family Health International to plan quality assurance programs in the countries of Thailand and the Philippines. Dr. Constantine has been author or co-author of over 40 scientific articles, 52 abstracts, and is the primary author of 2 books and 2 book chapters.

Johnny D. Callahan, M.T. (ASCP) is a full time Graduate Student/Research Assistant in the Department of Pathology at the University of Maryland School of Medicine at Baltimore.

He graduated from the Medical University of South Carolina, Charleston in 1987 with a B.S. degree in Medical Technology, is a registered Medical Technologist and is an associate member of the American Society of Clinical Pathologists (ASCP).

In the past, he was a staff Medical Technologist in the Diagnostic Immunology laboratory at the Medical University of South Carolina Hospital in Charleston from 1987 to 1988; and the supervisor of the Clinical Diagnostics Branch, Virology Division, at the U.S. Naval Medical Research Unit #3 in Cairo, Egypt from 1989 to 1990.

Mr. Callahan has acted as a WHO advisor and presented lectures at WHO-sponsored HIV workshops in Egypt and Kuwait; and was a WHO consultant and team member for the WHO Global Program on AIDS (GPA) laboratory component to help formulate a medium term plan for the control of HIV infection in Egypt.

Johnny D. Callahan is the co-author of a book on HIV testing, has published seven journal articles, and is co-author on six scientific abstracts. His main research interests include the development of rapid diagnostic assays for the human retroviruses.

Douglas M. Watts, Ph.D. is the Director of the Infectious Diseases Threat Assessment Research Program at the U.S. Naval Medical Research Institute, Bethesda, Maryland.

Dr. Watts received a B.A. degree in 1965 from Berea College, Berea, Kentucky, M.S. degree in 1972 and Ph.D. degree in 1974 from the University of Wisconsin, Madison. He served as chief of an Arboviral Diseases Research Program from 1974 to 1977 at the Walter Reed Army Institute of Research, Washington, D.C.; Assistant Chief of the Department of Virology from 1977 to 1980 at the U.S. Armed Forces Research Institute of Medical Sciences, Bangkok, Thailand; Senior Scientist of an Arboviral Diseases Research Program from 1980 to 1982 at the U.S. Army Medical Research Institute of Infectious Diseases (USAMRIID), Ft. Detrick, Frederick, Maryland; Medical Research Advisor for the World Health Organization (WHO) from 1982 to 1984 at the WHO Southeast Asia Regional Office, New Delhi, India; Senior Scientist of Virology Research Program from 1984 to 1988 at USAMRIID; Head of the Department of Virology from 1988 to 1991 at the U.S. Naval Medical Research Unit #3, Cairo, Egypt; and in 1991 assumed present position.

Dr. Watts is a member of the American Society of Tropical Medicine and Hygiene, American Association for the Advancement of Science, and the National Council for International Health. He has served as a guest lecturer at several national and international universities, a WHO consultant and advisor, member of U.S. Army ad hoc review committee, reviewer of scientific manuscripts and research proposals on tropical disease research for several scientific journals, and reviewer and advisor for the Board on Science and Technology for International Development, National Research Council, Washington, D.C.

Dr. Watts is the author or co-author of 70 scientific papers, including 1 book chapter and co-author of 1 book. His current major research interest relate to the diagnosis and epidemiology of viral diseases.

FOREWORD

Human immunodeficiency virus (HIV) infection and the acquired immunodeficiency syndrome (AIDS) are of major concern to health care professionals in all parts of the world. In addition, infection by the related retroviruses, HTLV-I and HTLV-II, is recognized as a significant public health threat. Now that testing for retroviral infections has been established in laboratories throughout the world and has been mandated in many countries, it is time for education to proceed beyond that of the mechanics for performing the tests. Assurance that the testing process will be efficient and accurate requires that laboratory personnel possess advanced knowledge in all aspects of the laboratory investigation. To minimize and avoid errors, it is essential that individuals who perform the tests have a thorough understanding in the proper operation of the laboratory and the techniques available for testing. This includes detailed knowledge in the proper means to handle and process specimens, the choice and use of tests, the interpretation of results, and the necessary quality assurance and quality control measures needed to effectively and accurately provide a correct laboratory diagnosis. Laboratory personnel must possess the background knowledge and technical expertise to be confident that results are as accurate as the limits of the tests permit. In addition, laboratory supervisors must be able to teach technologists and monitor their work to ensure not only effectiveness in the testing process and also continually update information. In order for laboratory personnel to produce quality results, they must have a basic understanding of the retroviruses and their biochemical composition, the immune responses generated by the host, and the basic principles of the diagnostic tests. This understanding is important for the proper selection, use, and interpretation of diagnostic tests.

The authors of this book have been trained in laboratory medicine and have strong backgrounds in immunology. Dr. Constantine has taught clinical immunology to medical technology students for over 8 years, has directed an HIV diagnostic and research laboratory in a developing country for almost 3 years, and currently directs a diagnostic immunology laboratory at the University of Maryland Medical Systems/Hospital (U.S.). Dr. Watts currently heads a division within the U.S. Navy that investigates infectious disease threat, and has been a World Health Organization (WHO) consultant and technical representative for medical research in developing countries. Johnny Callahan has worked in a clinical diagnostic laboratory for 5 years and has had experience supervising an HIV testing laboratory in a developing country for 2 years. Each author has trained technicians and technologists from many parts of the world, and each has acted as a WHO consultant or advisor in several countries for the laboratory diagnosis of HIV. In addition, all have taught HIV testing workshops in several countries over the past years, and therefore have developed insight into both the problems that laboratory personnel encounter, and areas in which a more thorough understanding is needed.

Our previous publication, entitled *HIV Testing and Quality Control: A Guide for Laboratory Personnel* (Family Health International, ISBN 0-939704-07-2) was prepared for use in developing countries. The current book has been updated, expanded to include the other retroviruses, and includes a more thorough explanation of all areas of retroviral testing and quality assurance.

The objectives of this book are (1) to provide basic information on retroviruses and retroviral testing; (2) to describe the available testing technologies and how these should be selected and used; (3) to provide information on how the test results should be interpreted; and (4) to describe the essential quality control and quality assurance measures that must be implemented to ensure that results are as accurate as possible.

This volume represents an effort to establish the capability for effective, precise, and accurate laboratory testing for infection by the retroviruses. In order to prevent any serious consequences from an incorrect retroviral test result, laboratory personnel must be well informed, technically competent, and must use the total of all pertinent information in order to provide a quality laboratory diagnosis.

Niel T. Constantine

PREFACE

As HIV infection has spread throughout the world, the need for HIV testing has increased. Many countries have well-developed laboratory approaches for protecting their blood supplies as well as for supporting diagnosis and medical referral of HIV-infected persons. Most of the HIV antibody screening and confirmatory testing in these countries is performed at the local level by many different types of laboratories, while centralized laboratories may perform cultures and other sophisticated tests, or provide reference diagnostic services for problematic sera. In contrast, laboratories in less developed countries must rely on simple, inexpensive screening tests performed at the local level, with limited reference service available through a central laboratory that may take samples 1 week or more to reach. Also, there are countries whose laboratory facilities fall between these extremes.

Regardless of the laboratory's level of involvement in HIV testing, the training and expertise of staff members are major determinants of the quality of test results produced. Performing a few simple tests very well is better than offering a wide array of tests with unreliable results. Laboratory personnel must strive for total quality assurance, seeing to it that every step of the testing process over which they have control is monitored and verified for accuracy and precision. The process begins with obtaining the proper specimen from the correct patient, and proceeds through preparing the specimen for testing, selecting the appropriate test, performing the test, validating its results, and reporting the correct result for the correct patient. The laboratory and the health care provider must cooperate before and after testing to ensure that the provider's diagnostic needs are met by the laboratory's testing menu, and that the results are interpreted accurately. Within the laboratory, personnel must feel confident that they are protected from injury so that they can do their best work safely, without fear and distraction.

Laboratories should have in place procedures that are followed by everyone, procedures that completely address each step of the testing process, including the safety precautions to be observed. Laboratories should ensure that their staff are thoroughly acquainted with these procedures, and that newly hired personnel are introduced to and trained in them. The entire laboratory has an investment in the accuracy of its results, and staff should regard every specimen, every test, and every result as if it were their own. Finally, laboratories should enroll in an external quality assessment, performance evaluation, or proficiency testing program in order to gain an objective assessment of their performance. Such programs are available through public health, national, or commercial laboratories and accrediting agencies.

Many books suitably address the principles of general laboratory testing. This book addresses those principles in the context of HIV testing, and has something to offer all laboratories and laboratorians, experienced as well as novice. For those new to HIV testing, it introduces the retroviruses, the immune system, and the variety of tests used to detect or monitor them. Chapters that relate to quality assurance and selection of tests will be valuable even to the experienced laboratorian. Since the book is intended for worldwide distribution, it also incorporates discussion of some tests currently not available in the U.S.; this information may help expand the U.S. reader's knowledge of a variety of unfamiliar test formats.

Retroviral Testing: Essentials for Quality Control and Laboratory Diagnosis provides important insights into difficulties that may be encountered during laboratory testing for HIV. This book should stimulate laboratory personnel to examine their testing and quality assurance protocols for improvements that may be needed. Although not all the answers to every testing challenge will be found here, reasonable approaches may be suggested by the information provided. The reader should expect nothing less.

Dr. Wanda K. Jones
Centers for Disease Control
Atlanta, Georgia

ACKNOWLEDGMENTS

The authors wish to acknowledge the contributions of the following persons, who provided a thorough and thoughtful technical and editorial review of the book: Dr. Wanda K. Jones, CDC, Atlanta, GA, U.S.; and Robert Myers, Department of Health and Mental Hygiene, State of Maryland.

Appreciation is also extended to Johnny Callahan for his expertise in producing the illustrations; to Dr. Xiang Zhang, University of Maryland School of Medicine (U.S.), and our families, for their assistance and support. We also wish to recognize the important contributions to the earlier version of this book by Sheila Mitchell, Robert Gringle, Dr. Edwin Archbold, Dr. Peter Lamptey, and Debra Hanson of AIDSTECH/Family Health International, Durham, NC, U.S.; Dr. Guido van der Groen of the Institute of Tropical Medicine, Antwerp, Belgium; Dr. Hiko Tamashiro of the Global Programme on AIDS, World Health Organization, Geneva, Switzerland; and Dr. Debrework Zewdie, National Research Institute of Health, Addis Ababa, Ethiopia.

TABLE OF CONTENTS

Introduction to Retrovirology and
Retroviral Testing 1

Chapter 1
HIV-1: Biology 5
I.   The HIV-1 Virus 5
II.   HIV Infection and AIDS..7
III. HIV-1 Antigens 10
References 13

Chapter 2
The Immune System during HIV-1 Infection 15
I.   Antibody Responses 15
     A.   Adults 15
     B.   Newborns 19
II.   Cellular Considerations during Infection 20
     A.   Lymphocytes, T Cell Subpopulations,
          Cell Interactions 20
     B.   Cytokines 23
     C.   Laboratory Investigation of Cellular Abnormalities................25
          1.   Cytokine Assays 25
          2.   Methods to Quantitate T and B Lymphocytes 26
          3.   Normal Values 28
          4.   Expected Values during HIV Infection
              and AIDS 30
     D.   Functional Defects of Immune Cells 31
     E.   Other Markers for Cell-Mediated Immunity
          during Infection 32
References 32

Chapter 3
Screening Tests for HIV-1 Infection 35
I.   Introduction 35
II.   Principles of the Tests 37
     A.   ELISAs 37
          1.   Indirect 39
          2.   Competitive 39
          3.   Antigen Sandwich 40
          4.   Antigen and Antibody Capture......................................41
     B.   Agglutination Tests 42
     C.   Dot-Blot Assays 46

	D.	Other Types of Screening Assays 47
	E.	Recombinant and Synthetic Peptide Antigen-Based Tests 48
III.	Urine and Saliva Tests 50
IV.	Interpretation of Screening Test Results 51
V.	Selection of a Screening Test 53
VI.	Commercially Available Screening Tests 54
References 57

Chapter 4
Supplemental Tests for HIV-1 Infection 59
I. Introduction and Use 59
II. Principles of Tests 60
    A. Western Blot 60
    B. Line Immunoassays (LIA) 64
    C. Indirect Immunofluorescence Assay (IFA) 65
    D. Radioimmunoprecipitation Assay (RIPA) 67
    E. Tests to Detect Circulating Antigen 68
    F. *In Vitro* Antibody Synthesis 70
    G. Viral Culture 71
    H. Assays to Detect Viral Nucleic Acids 73
III. Interpretation of Confirmatory Test Results 74
    A. Criteria for Positivity 75
        1. Western Blot Criteria 75
        2. Indirect Immunofluorescent Assay Criteria 77
    B. Inconclusive Results 79
IV. Advantages and Disadvantages of Confirmatory Tests 80
V. Testing Algorithms 82
VI. Commercially Available Confirmatory Assays for HIV 84
References 85

Chapter 5
HIV-2 89
I. Introduction 89
II. Antigens of HIV-2 90
III. Testing for HIV-2 91
    A. Screening Tests 91
    B. Confirmatory Tests 94
    C. HIV-1/HIV-2 Combination Assays 95
        1. Combination Screening Tests 95
        2. Combination Confirmatory Tests 97
    D. Interpretation of Test Results 100
References 101

Chapter 6
HTLV-I and HTLV-II 103
- I. Introduction 103
- II. Disease Associations 103
  - A. HTLV-I 103
  - B. HTLV-II 104
- III. Antigens 104
  - A. HTLV-I 104
  - B. HTLV-II 105
- IV. Screening Tests for HTLV-I/-II 105
- V. Confirmatory Tests and Differentiation of HTLV-I and HTLV-II 107
- References 110

Chapter 7
Indicators of the Value of Diagnostic Tests 113
- I. Introduction 113
- II. Sensitivity 113
- III. Specificity 114
- IV. Test Efficiency 115
- V. Delta Values 115
- VI. Predictive Values 117
- References 118

Chapter 8
Quality Control and Quality Assurance 121
- I. Necessity and Importance 121
- II. Confidentiality 122
- III. Definitions 122
  - A. Quality Control 122
  - B. Quality Assurance 122
  - C. Quality Assessment 122
- IV. Quality Assurance: Fundamentals for Overall Quality of Results 122
  - A. Recordkeeping 123
  - B. Monitoring Laboratory Staff 124
  - C. Vigilance in the Laboratory 125
  - D. Verification of True Positives and True Negatives 125
  - E. Parallel Testing of Resubmitted Specimens 126
  - F. Reviewing Transcriptional Measures 126
  - G. Reporting of Results 126
  - H. Interaction with Physicians 127
  - I. Storage of Specimens for Follow-Up Testing 127
  - J. Laboratory Efficiency 129
  - K. Total Quality Management 130

		1.	Introduction 130
		2.	Evaluation of Laboratory Staff... 131
		3.	Continuing Education 132
		4.	The Laboratory as a Diagnostic System 132

- V. Quality Control: Monitoring the Testing Process 134
  - A. Internal and External Controls 135
  - B. How to Determine Acceptance of Control Values 137
  - C. Calculations 138
    1. Mean 138
    2. Standard Deviation 139
    3. Coefficient of Variation 140
  - D. Plotting Quality Control Graphs 143
  - E. Use of Calculated Values 145
  - F. Shifts and Trends 148
  - G. An Example of a Quality Control Protocol 151
  - H. Use of Duplicate Controls 152
  - I. Recalculation of Acceptable Control Values 153
    1. Changes in the External Control Lot 154
    2. Changes in Test Kit Lot 154
  - J. Calculation of "Gray-Zone" Reactors 155
  - K. Use of Gray-Zone Results 155
- VI. Quality Assessment: Monitoring Laboratory Performance 156
  - A. Proficiency Panels and Blind Testing 156
- VII. Equipment Maintenance and Calibration 158
- VIII. Common Errors Encountered during HIV Testing 160
- IX. Troubleshooting 161
  - A. General Assay Problems 161
  - B. Specific Assay Problems 162
  - C. Instruments 164
  - D. Internal Control Problems 165
  - E. External Control Problems 165

References 166

Chapter 9
Laboratory Techniques 169
- I. Introduction 169
- II. Specimen Collection 169
  - A. Types of Specimens 169
  - B. Blood Collection Methods 170
    1. Syringes 170
    2. Vacuum Tube Devices 171
    3. Filter Paper as a Carrier for Blood Specimens 171
- III. Specimen Processing 172
  - A. Receipt of Specimens 172

	B.	Acceptable Specimens 172
	C.	Clotting and Centrifugation 172
	D.	Pooling of Specimens for HIV Testing 172
IV.	Volumetric Measurements 173	
	A.	Pipettes 173
		1. Single-Channel Manual Pipettes 175
		2. Multichannel Manual Pipettes 176
	B.	Mechanical and Motorized Pipetting Aids 176
	C.	Pipetting Technique 176
		1. Serological Pipettes 176
		2. Manual Pipettes 176
	D.	Reagent Reconstitution and Glassware 178
V.	Preparing Dilutions 179	
	A.	Simple Dilutions 179
	B.	Serial Dilution Technique 180
VI.	The Testing of Preparations Intended for Human Injection 183	
	A.	Dilution of Concentrated Products 185

References 186

Chapter 10
How to Evaluate Diagnostic Test Kits 187
I. Introduction 187
II. Reference Tests 187
III. Choice of Specimens 188
IV. Testing Conditions 189
V. Discrepant Results between Assays 189
References 190

Chapter 11
Safety in the Laboratory 191
I. Introduction 191
II. Occupational Risks 191
    A. Infectious Material 191
    B. Accidental Exposure 192
    C. Work Space 192
    D. Biohazardous Waste 193
III. Universal Precautions 194
References 195

Summary and Closing Remarks 197

Glossary 199

Appendices 213

Index 221

INTRODUCTION TO RETROVIROLOGY AND RETROVIRAL TESTING

Retroviruses are the cause of considerable human morbidity and mortality. Although these agents have been known to exist and cause disease in animals for nearly 1 century, their role as human pathogens has only been recognized within the last 15 years. With the development of sensitive methods such as the reverse transcriptase assay in the 1970s and the ability to culture human T lymphocytes, isolation and identification of the retroviruses as human pathogens became possible. The four human retroviruses, human immunodeficiency virus types 1 and 2 (HIV-1, HIV-2) and human T lymphotropic virus types I and II (HTLV-I, HTLV-II), share many common features, but are distinct viruses. Among their similarities are their modes of transmission, tendency to cause latent infections (usually requiring many years to cause disease), and their CD4+ cell tropism. Most dramatically, they differ in the diseases they produce: immunodeficiency by the HIVs and immunoproliferation by the HTLVs. The HIVs are cytolytic viruses, while the HTLVs are transforming viruses.

The importance of retroviral infections is evident, as reflected by current statistics. Worldwide, the number of reported cases of acquired immunodeficiency syndrome (AIDS) caused by the HIVs is currently approaching 500,000, with fatal results in nearly 65% of cases. However, delays and reluctance in accurately reporting cases makes this number suspicious. It has been estimated that the actual number of individuals with AIDS currently exceeds 1 million. More alarmingly, it is believed that there are between 8 and 12 million persons presently infected worldwide with the HIV viruses, all of whom will eventually die of AIDS. By the year 2000, it is estimated that 40 million persons may be infected with HIV. In the U.S., AIDS is the second leading cause of death among men 25 to 44 years of age and will soon be one of the five leading causes of death in women aged 15 to 44. AIDS has become the leading cause of death for women aged 20 to 40 in major cities of the Americas, Western Europe, and sub-Saharan Africa. Some countries in Central Africa have reported that the prevalence of HIV infection is approaching 30% of the urban adult population. A recent report from the World Health Organization (WHO) estimates that one half of all AIDS cases occur in Africa, but the largest rate of yearly increase in cases is occurring in some countries of Asia. In one Asian country, it has been estimated that up to 95% of donated blood is unsafe, with 80% of one city's blood sellers being HIV infected. Also of considerable concern is the recent increase in the cases of HIV infection and AIDS in children throughout the world. In the 1990s AIDS will be a leading cause of morbidity and mortality in children in the U.S. For each of the children diagnosed with AIDS, it is estimated that 5 to 10 more are infected with HIV. Worldwide, WHO estimates that 1 million children are infected with HIV.

The importance and epidemiology of infection associated with the HTLVs have only been recognized since the 1980s. Endemic areas of HTLV-I infection include Japan; the Caribbean islands; the southeastern U.S.; and parts of South America, Italy, and Africa. The prevalence of HTLV-I infection is reported to be 5 to 10% in some areas of Japan. Little information is presently available on the epidemiology of HTLV-II infection, although very high rates have been identified in intravenous (i.v.) drug addicts in the U.S. The routes of transmission of the HTLVs are similar to those of the HIVs, namely via blood transfusion, vertically from mother to fetus, and sexually. Blood transfusion is the most efficient mode, with reported seroconversion rates of 35 to 60% following exposure to contaminated blood. Infection by the HTLVs has been associated with a variety of diseases including adult T cell leukemia/lymphoma (ATLL), HTLV-I associated myelopathy/tropical spastic paraparesis (HAM/TSP), and hairy cell leukemia.

The worldwide institution of laboratory testing for the retroviruses, in particular for HIV, has had a major impact in helping to contain the pandemic. AIDS (formerly GRID, gay-related immune deficiency) has only been recognized as a disease since 1982, although "gay cancer" was reported by *Morbidity and Mortality Weekly Report* in mid-1981, and only within the last 7 years has a laboratory diagnostic test for HIV infection become available. Tests to detect infection by the retroviruses (HIV-1 and HTLV-I/II) are now included among the other tests required by the U.S. Food and Drug Administration (FDA) for screening blood prior to transfusion. In 1992, the FDA has recommended that blood be screened for HIV-2 antibodies. Retroviral diagnostic tests have been instrumental in helping to protect the blood supply, the surveillance of populations to define disease threat, and in identifying infected individuals so that better health care can be provided. The detection of antibodies in serum is still the most widely used method for the laboratory diagnosis of retroviral infections.

Presently, well over 100 different diagnostic assays from more than 40 commercial companies are available to detect antibodies to the retroviruses (Appendix A). Although the enzyme-linked immunosorbent assay (ELISA) and Western blot (WB) assay are still the most popular, alternatives exist which offer a choice for different testing situations. However, local regulations may permit only certain tests to be used; for example, in the U.S. only FDA-licensed screening tests can be used. The tests currently in use in most countries are sensitive and specific, but are not free from false-positive and -negative results. Therefore, supplemental, confirmatory, or alternative tests must be used. However, even with the best tests, the quality of results critically depends on the proficiency of the individuals who perform the tests.

Until effective prevention and treatment strategies are developed, and universally implemented, the prevalence of HIV infection will continue to increase exponentially. For the foreseeable future, laboratory diagnosis will continue to play a significant role in identifying HIV-infected individuals so that they can be counseled and educated in an effort to prevent the infection of others. The serious ethical, legal, and social issues that accompany HIV infection demand that test results be as accurate as possible.

REFERENCES

Chin, J., Current and future dimensions of the HIV/AIDS pandemic in women and children, *Lancet*, 336, 221, 1990.
Kandela, P., India: HIV banks, *Lancet*, 338, 436, 1991.
CDC, The HIV/AIDS epidemic: the first 10 years, report 7, *MMWR*, 357, 1991.
AIDS Administration of the Maryland Department of Health & Mental Hygiene, Early identification of HIV infection in children, *Communicable Dis. Bull.*, October 1991.

Chapter 1

HIV-1: BIOLOGY

I. THE HIV-1 VIRUS

The HIV-1 virus is an icosahedral, enveloped, RNA virus of the Lentivirinae subfamily of retroviruses that primarily infects human white blood cells. They are so named because their genetic material, RNA, must be transcribed to DNA before the virus can complete its replicative cycle. The transcription of viral RNA into single-stranded DNA is accomplished by an enzyme, reverse transcriptase (RT). After the synthesis of a second strand of DNA by host cell mechanisms, the DNA is integrated (inserted) into the host cell DNA. The host cell DNA, containing the viral genome, is now considered to harbor the HIV provirus.

The proviral DNA is transcribed into messenger RNA (mRNA) and the resultant RNA is translated along with that of the host's DNA as part of the host's normal cellular processes (Figure 1). A portion of the viral genome of the provirus, the long terminal repeat (LTR), is recognized by enzymes that bring about the synthesis of specific viral components through the host cell's translational processes. The resultant components, proteins and glycoproteins, eventually assemble to produce complete viral particles that are released as infectious virions by budding from the host cells. However, active replication of HIV may not always be accomplished, and a long latency period may occur after infection. Following some unknown event occurring usually years later, the replicative cycle resumes, resulting in excessive viral production.

The genome of HIV contains two types of genes: the structural genes and the regulatory genes. The structural genes are responsible for the direction and synthesis of proteins and glycoproteins that will give the virus its physical characteristics; i.e., shape, size, structural integrity, compartmentalization, etc. The regulatory genes are responsible for the subsequent production of proteins that can affect the activities of viral components, or can specifically turn other genes on and off. Among other activities, the regulatory proteins have the ability to increase or decrease the replication of HIV. It is probable that the activity of these regulatory genes is responsible for the profound pathogenicity of HIV.

Structurally, HIV is composed of two major parts: an outer envelope and an inner core (Figure 2). The external envelope components are embedded in the lipid matrix of the membrane and are involved in the binding of the virus to the host cells during infection. The outermost viral components are arranged as spikes or knobs and extend from the lipid membrane. These external spikes are proteins with sugar molecules and are therefore glycoproteins. The core components of the virus are located internal to the outer membrane and are bound by a protein coat encompassing two identical copies of the nucleic acid RNA

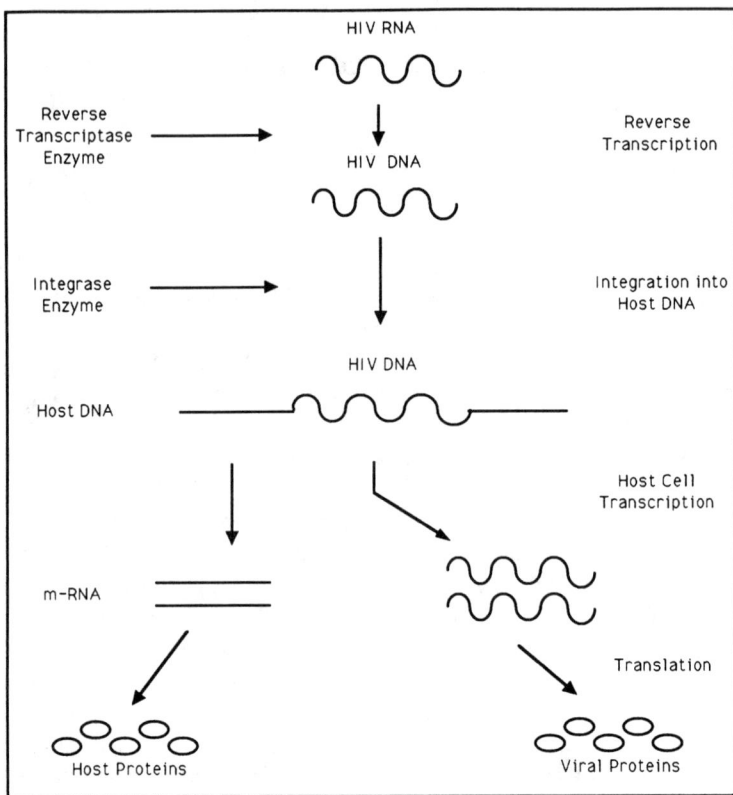

FIGURE 1. Translation, integration, and transcription of the viral genome.

genome. Also within the core are three viral enzymes: RT, integrase, and protease. These enzymes are responsible for transcription, integration of the virus in the host, and cleavage of other proteins, respectively. The specific components of the envelope and core are described below in Section III, "HIV-1 Antigens".

Very importantly, some of the structural components (mainly the envelope components) become modified throughout infection, probably as a result of mutation and/or immunologic selection. These changes may prevent the immune response of the host from being effective in containing the infection. Therefore, small differences in the precise amino acid sequences of the HIV envelope components may occur between different individuals, and even within the same individual during different stages of the infection. This is one of the reasons that an effective vaccine may be difficult to develop.

HIV, as with most enveloped viruses, cannot survive for very long unless it is contained within cells; it must maintain an intracellular relationship. Therefore, infectious viral particles are mostly present in fluids that contain

HIV-1: Biology 7

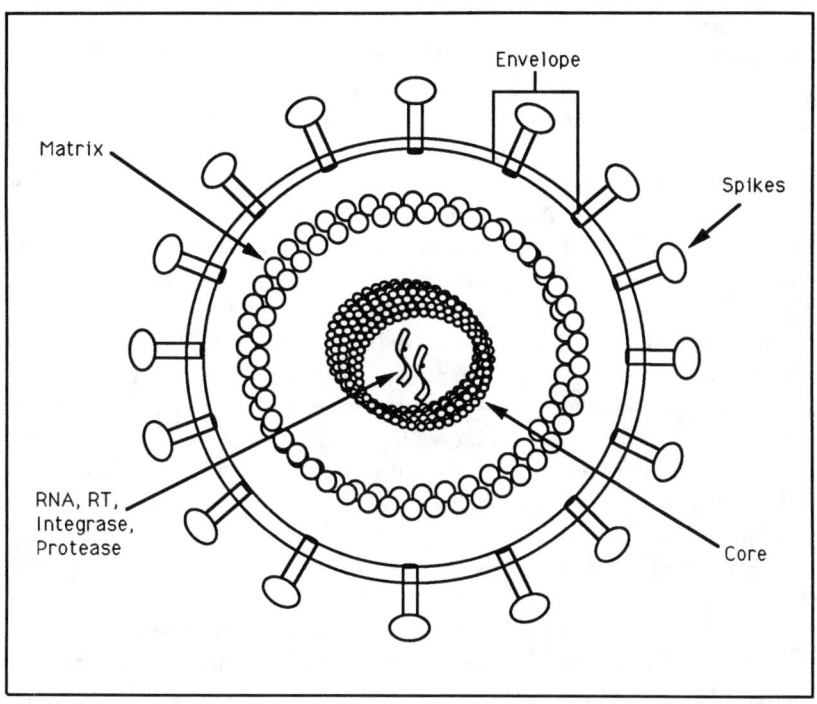

FIGURE 2. The major structural components of HIV.

infected cells, although virus also has been associated with cell-free fractions of blood. Components of the virus (soluble antigens) that are not infectious may also be detected in cell-free fluids such as serum, indicating that the individual has been infected.

II. HIV INFECTION AND AIDS

The events leading to viral attachment, penetration, replication, and pathology are complex and will not be addressed in detail. The processes are the same for both HIV viruses. Basically, the virus attaches to the host cell (most often a lymphocyte) through a complicated mechanism involving the external envelope components of the virus and a protein on the host cell. Since the virus can attach to this host protein, the protein can be considered to be a receptor for the virus. This receptor is known as CD4 and is the same molecule that identifies these cells as a subpopulation of lymphocytes called helper T lymphocytes or CD4+ T cells. The letters CD indicate cluster differentiation, and represent a population of cells as they mature and differentiate into a specific cell line during ontogeny. Many other types of human cells (e.g., macrophages and glial cells of the brain) may have this CD4 receptor (or one similar), and therefore also can be infected by the virus. Evidence is accumulating to indicate that other body cells that do not have this CD4 receptor can be infected by HIV, presumably via a different mechanism. In contrast to infection of the CD4+ T cells, infection of the monocytic cell line does not lead readily to an apparent cytopathic effect, and virus production may not occur via budding but by membranes within intracytoplasmic vesicles. In any event, the virus fuses with the cell, internalizes, and transcription and translation ensue. At this stage, the person is infected with HIV. Using new and very sensitive techniques it has been suggested that 1 in every 100 to 10,000 CD4+T cells of asymptomatic individuals may be infected; 1 in 100 cells of patients with AIDS.

Soon after entry into the cell, a transient replicative cycle commences as the viral genome (RNA) is transcribed and integrated into DNA by the viral enzyme RT and the host enzyme ribonuclease H. At this time there are few, if any, symptoms in the infected individual. The eventual production of viral components through translation, assembly, and subsequent release of infectious virions by budding results in the infection of neighboring cells (Figure 3). It is during these early stages of infection (within weeks) that antibodies to the viral components are produced and can be detected in the serum of the infected individual.

As a result of infection, some host cells are destroyed, some fuse together to form masses of cells known as syncytia, and some simply exist, seemingly unaffected (latent infection). Direct destruction of host cells probably occurs due to

1. The lysis of cells and resultant ionic imbalances when the virions bud from the cell after viral replication
2. Damage to the intracellular components of the host cells by the antigens and viral particles
3. Syncytia formation, which involves uninfected cells
4. Autoimmune phenomena involving the CD4 molecule, which may have structural similarities with the class II Major Histocompatibility Complex (MHC) molecule on human lymphocytes
5. Antibody-dependent cytotoxicity, in which bystander CD4+T cells bind viral antigens
6. The possible infection and destruction of precursor cells of the CD4+T lineage

Disease symptoms usually do not exist at the time of HIV infection, and may not appear for many years (asymptomatic infection). For a short time following infection, viral proliferation is controlled, most likely by the immune responses of the host. The latency period ensues, during which time the virus is dormant (virus replication is slow). Years later (an average of 10 years elapses before symptoms appear), and after some unknown event, the virus (provirus) is activated and replication occurs at a high level. It is at this time that symptoms usually appear and the individual may be diagnosed as having

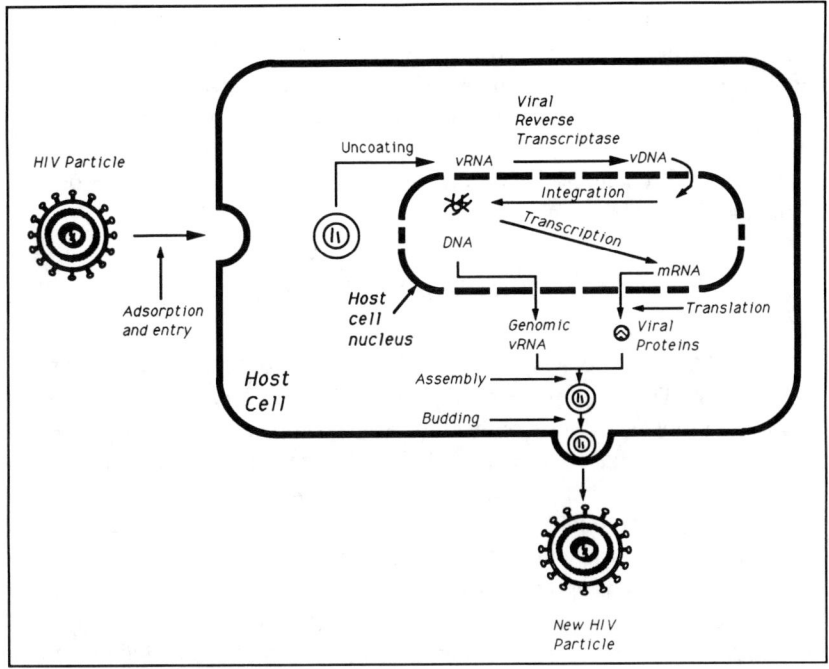

FIGURE 3. The HIV life cycle.

AIDS-related complex (ARC) or, if certain symptoms appear, full-blown AIDS. A number of factors may influence virus expression and the progression of disease. These may include mitogens, heterologous viruses, antigens, physical factors, and cytokines (Chapter 2).

Symptoms associated with AIDS may be diverse and may vary because of the involvement of many organ systems. The most frequently reported causes of death in persons afflicted with AIDS are due to infectious diseases caused by the ultimate destruction of immune cells, and the manifestation of immunoincompetence, the inadequate functioning of the immune system. The infectious diseases are usually caused by opportunistic pathogens, organisms that rarely cause disease in an immunocompetent individual. At the time the virus is activated and symptoms appear, a precipitous decline usually occurs in the number of CD4+ T cells. A direct correlation between viral burden, decreasing CD4+ T cells, and disease progression has been demonstrated. Because the helper CD4+ T lymphocytes orchestrate the activities and functions of most other immune cells, the effects of the virus on these lymphocytes can interfere with the functioning of other immune cells. When the relationship between the immune cells is lost, the entire immune system becomes ineffective.

Currently, individuals are considered infected with HIV when they have specific antibody identified by screening and supplemental HIV tests, or other methods, including direct identification of virus in host tissues by virus isolation, antigen detection, or detection of HIV genetic material (DNA or RNA). AIDS is considered the diagnosis for patients who are HIV infected and have evidence of severe immunosuppression, or a variety of opportunistic infections. According to the Centers for Disease Control (CDC) surveillance case definition of AIDS, severe immunosuppression is defined as an absolute CD4+ T lymphocyte count of $<200/mm^3$, or a CD4+ T cell percentage of total lymphocytes <14 if the absolute count is not available.

III. HIV-1 ANTIGENS

As with all infectious agents, the HIVs are composed of basic biochemical components. These components combine and arrange themselves to produce a complete virion, capable of replication. Some components function in a structural capacity, giving HIV its physical integrity and morphology, while others function to regulate the activities of the virus (enzyme expression, degree of replication, gene product formation, etc.). The components of HIV that are important for the production of the immune response include proteins (chains of amino acids); glycoproteins (proteins coupled to carbohydrates or sugars); and nucleic acids (complex combinations of base pairs that form the genes of the virus). Enzymes of the virus are also important, and are included with the protein group. All of these components are foreign when introduced into the human host and can react with the subsequent products of the immune response; hence, they are referred to as antigens. Many antigens are immunogenic and can elicit an immune response (in addition to their reaction with the products of the immune response; see Chapter 2). Antigens, whether structural or regulatory components, are produced as a result of instruction specified in the nucleic acids that make up the genome of the virus, and therefore these antigenic components may also be referred to as gene products as they are produced from instruction by the genes. As the virus replicates in the host, these viral antigens are often released into the blood of the individual. Therefore, at certain times during HIV infection antigenemia (soluble antigens in the blood) occurs in the host. The presence of antigens in the serum of an individual demonstrates that infection has occurred, can be an important means to monitor the progression of the disease, and can assist in the diagnosis of infections in newborns (Chapter 2).

The different structural antigens of HIV-1 are encoded by three major structural genes, and therefore can be classified into three major groups:

1. gag proteins (encoded by the *gag* gene, or group-associated antigen gene)
2. env glycoproteins (encoded by the genes that determine the production of the envelope components)
3. pol proteins (polymerase components that represent enzymes that are involved in transcription, integration, and cleavage of other viral components)

HIV-1: Biology

The most important gag proteins (p) are those having molecular weights of 55,000 (p55), 24,000 (p24), 17,000 or 18,000 (p17), and 15,000 Da (p15). The p55 antigen is a precursor molecule produced early in the infectious process and is eventually cleaved to produce the other core proteins. All of the gag proteins are located in the nucleocapsid of the virus. The p17 protein lies in the matrix between the protein core and the envelope and is embedded within the internal portion of the envelope. The p24 and p15 proteins make up the core coat (capsid) that surrounds the internal nucleic acids. The major capsid component is p24. In addition, recent reports using high resolution electron microscopy have revealed two additional core proteins, p9 and p7, which are nucleic acid binding proteins. Together these proteins, which are closely associated with the genomic RNA, make up the nucleoid core.

The env (envelope) antigens are glycoproteins and include those components having molecular weights of 160,000 (gp160), 120,000 (gp120), and 41,000 Da (gp41). The gp160 is the precursor molecule and is not a structural component (similar to the gag p55); it is a component that is produced during infection, but is later cleaved to form gp120 and gp41 (the structural envelope glycoproteins). These envelope glycoproteins contain:

- Conserved regions (sequences constant in all HIV-1 viruses)
- Variable regions (sequences that may vary between HIV-1 viruses)

The gp41 antigen spans the inner and outer membrane of the virus, and thus is often referred to as the transmembrane glycoprotein antigen. This particular antigen contains important variable regions, and therefore may be specific for each type or strain of the HIV-1 virus. The gp120 antigen is the major component of the external envelope and is responsible for the 72 knobs or spikes of the envelope. Significant variability in gp120 (as high as 15%) between the HIVs occurs in the hypervariable regions of the molecule; this may be responsible for the inability of the immune system to contain the virus. Together, the gp41 and gp120 antigens are involved with the fusion and attachment of HIV to the CD4 molecule on the host cells. The identification of antibodies to the envelope antigens is extremely important for the laboratory diagnosis of HIV infection. Antibody to at least one env component must be detected in the host in order to confirm infection (Chapter 4).

Polymerase (pol) antigens include the proteins p66 (a subunit of the reverse transcriptase enzyme that has RNase H activity), p51 (another subunit of the reverse transcriptase enzyme), and p31 (integrase or endonuclease). Polymerase antigens are located within the core of the virus and are closely associated with the nucleic acids. Collectively, these polymerase components (enzymes) are responsible for:

- Conversion of viral RNA to DNA (reverse transcription)
- Integration of viral DNA into host cellular DNA
- Cleavage of precursor molecules into smaller active components; accomplished by the proteases

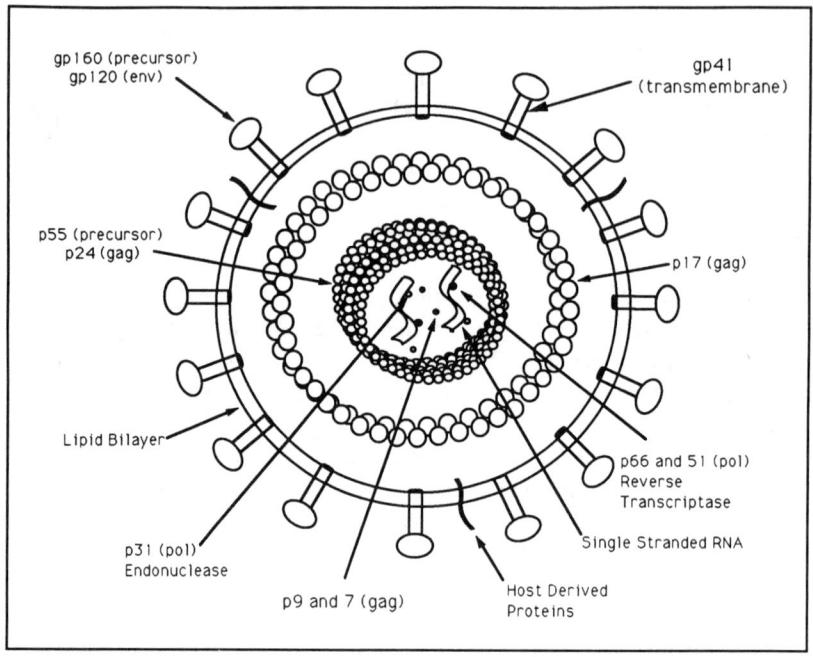

FIGURE 4. HIV-1 antigens.

Figure 4 depicts the location of gag, env, and pol antigens within the virus. The precursor molecules gp160 and p55 are not structural components of the virus and are only included in the figure for completeness. These precursor components are, however, true gene products produced during transcription and translation. It is later, but prior to virus assembly, that they are cleaved to form the other structural components. Therefore, during viral replication, these precursor molecules may be exuded into the blood and elicit antibodies that can be detected by serologic techniques.

Other components of the virus may be antigenic and play a significant role in infection and the development of immune responses, but their exact roles are poorly understood. These viral components include the negative regulatory proteins nef (p27) and vpr (p15), both of which may limit viral replication; and the positive regulatory proteins vpu (p16, increases the maturation of the viral particles), tat (p14, transactivates gene expression), rev (p19, production of viral mRNA), and vif (p23, viral infectivity factor). Generally, these regulatory components are responsible for modifying the expression of viral proteins and for the replication of the virus. The negative regulatory factors may be responsible for limiting the degree of viral replication (down regulation), thereby leading to the latency period. It may be that the event that ultimately activates the active replication of the virus is the inhibition of the negative regulatory factors. Although the exact function of these regulatory factors is unknown,

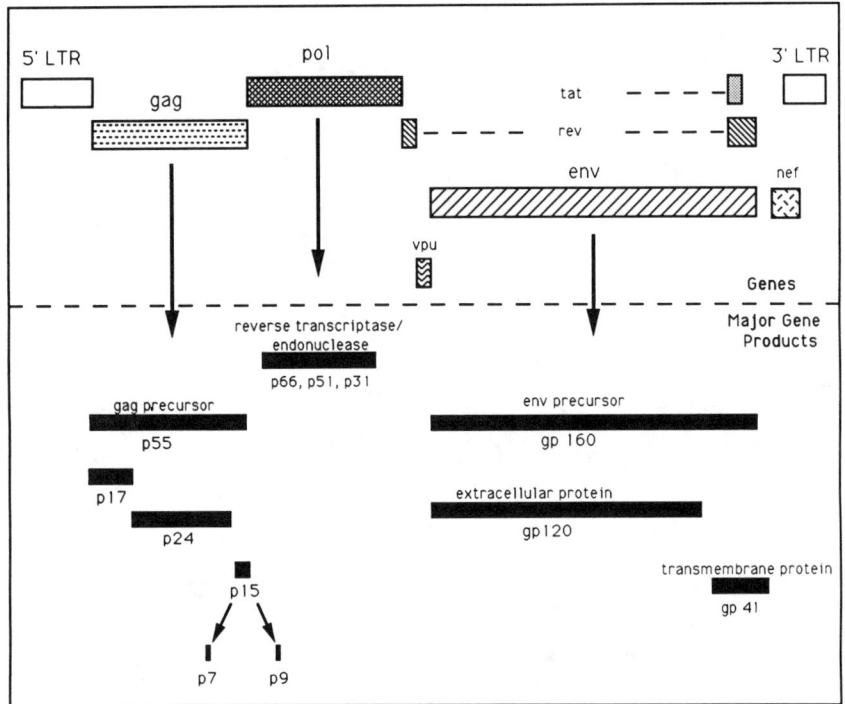

FIGURE 5. The HIV-1 genome and gene products.

they play an important role in the infectious process and may govern the whole process of disease progression. Consequently, the regulatory proteins are being investigated as targets for the development of antiviral therapies. Tests have been developed recently that can detect antibodies to some of these regulatory proteins. Although they are not used routinely at present, they may prove to be valuable in the future. Figure 5 depicts the relationship between the HIV genome and the gene products.

REFERENCES

Gallo, R. C., Human retroviruses: a decade of discovery and link with human disease, *J. Infect. Dis.*, 164, 235, 1991.

Green, W. C., The molecular biology of human immunodeficiency virus type 1 infection, *N. Engl. J. Med.*, 324, 308, 1991.

Kieny, M. P., Structure and regulation of the human AIDS virus, *J. AIDS*, 3, 395, 1990.

Rosenberg, Z. F., Immunopathogenesis of HIV infection, *FASEB J.*, 5, 2382, 1991.

U.S. Department of Health and Human Services, Public Health Service, Centers for Disease Control, 1992 Revised Classification System for HIV Infection and Expanded AIDS Surveillance Case Definition for Adolescents and Adults, draft, November 15, 1991.

Chapter 2

THE IMMUNE SYSTEM DURING HIV-1 INFECTION

I. ANTIBODY RESPONSES

A. Adults

As with most infectious agents, the human body responds to HIV infection by producing antibodies. These antibodies are usually produced between 6 and 12 weeks following infection, although in rare cases they may not be detected for months or years. If individuals are tested soon after infection they may not have detectable antibodies and will test negative. The individual is truly infected at the time of testing, but because antibody will not be detected, a negative result will be obtained. Therefore, it is important to understand the kinetics of the immune response in order to recognize when an individual should or should not be tested. However, a test should be performed in the early stages after exposure in order to produce a baseline result for later comparisons. Therefore, a negative result cannot be used as an indicator that the individual is absolutely not infected. As the concentration (titer) of antibodies gradually rises (as infection progresses), the antibody test result will change from negative to positive (seroconversion), and infection will be verified. Some antibodies to HIV components will persist throughout the infection, while other antibodies will decrease to undetectable levels. Specific antibodies are produced to all of the major antigens of the virus (p15, p17, p24, p31, gp41, p51, p55, p66, gp120, gp160), and to some of the regulatory proteins.

Although not useful for diagnostic purposes, antibodies that are capable of neutralizing the virus (neutralizing antibodies) are usually produced during infection. These antibodies contribute to the removal of the offending virus by at least two mechanisms. They can bind to viral antigens or to antigens expressed on virus-infected cells. Subsequent recognition of the resultant immune complexes by other immune cells serves as a basis for clearance, perhaps by antibody-dependent cellular cytotoxicity (ADCC). Also, neutralizing antibodies can bind directly to the virus, preventing the attachment of virus to CD4+ cells. Neutralizing antibodies are produced early during infection and may decrease precipitously in the symptomatic stages of HIV infection. These antibodies are most often produced against specific regions of gp120. A discrete area of the gp120 molecule known as the V3 loop (variable region) is one of the target regions. Although neutralizing antibodies theoretically should slow the infection by binding to and inhibiting the infection of other cells, they do not. No direct correlation seems to exist between the presence of neutralizing antibodies and disease state or prognosis. It may be that neutralizing antibodies become useless as the virus changes its antigenic components (envelope antigens). Nevertheless, much interest abounds in these antibodies and their effects during HIV infection and AIDS.

It is important to note that not all individuals will have identical immune responses to the virus. Some persons may have stronger responses to p24, for example, while others may respond strongly to other antigens. Although it is not unusual to demonstrate different antibody profiles in the sera from different individuals who are infected, most individuals respond to all viral components during the course of infection. The concentration of specific antibody to any or all antigens can also vary between persons, and vary within the same individual during various times of infection.

Antibody titers are commonly as high as 50,000, but can also be very low. Low titers are present early in infection (during the seroconversion period) and later in disease when the individual becomes immunodeficient. Low titers may be responsible for weak reactions in serological assays and can result in test false-negative results if the test lacks sensitivity (Chapter 7). In addition, the demonstration of specific antibodies may depend on the method used to detect these antibodies. For example, if a test primarily detects antibodies to gp41, it may not produce a positive result if the individual only has antibodies to p24 and gp120. Therefore, it is important to understand not only the kinetics of the response, but also the basis of the tests that are used. As discussed in Chapter 8, seroconversion can be demonstrated early if initial and repeat specimens are tested in parallel. Although a definitive diagnosis cannot be made unless the criteria for positivity are met, a rise in titer of antibody between the two samples can suggest that infection has occurred. Similarly, qualitative changes (increases in the number of viral-specific antibodies) may indicate that a person is truly infected.

Figure 6 illustrates the typical antibody response and the occurrence of antigenemia following infection by HIV. As indicated, antibodies to the major gag proteins (p55, p24) usually appear early during infection, while antibodies to the env and pol products are produced at the same time or slightly later. However, it is important to realize that the detection of these particular antibodies depends on the assay used; e.g., tests designed to detect primarily gag components may yield reactive results before tests that are designed to detect env antibodies. Similarly, some tests may detect gp41 antibody prior to the detection of gag antibodies.

Antibodies to p24 may decrease in time as the disease progresses, and as viral protein (p24 antigen) increases in the serum (Figure 6). This reversal of antibody/antigen most likely reflects sequestration and removal of antibody as it combines with an increasing amount of viral antigen produced as a result of virus replication. Alternatively, a decrease in antibody production may occur as the immune system becomes less responsive. Viral replication usually occurs late in the disease and is accompanied by disease symptoms (i.e., the AIDS syndrome becomes evident).

The appearance of p24 antigen in the serum (antigenemia) very early after infection is also to be noted. Although this may occur, the quantities of antigen may be very low, and may not be detected by antigen assay methods (Chapter 4).

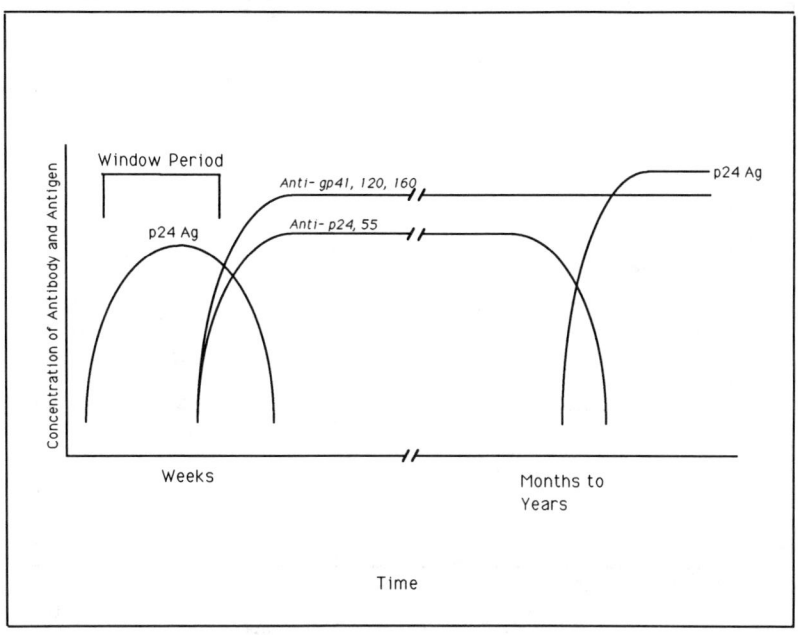

FIGURE 6. Antibody response and antigenemia during HIV infection.

The period after infection has occurred but before antibody can be demonstrated (generally 0 to 6 weeks), is called the "window period". It is during this window period that a test for antibody will be negative, although the individual is indeed infected. Therefore, the antibody test will produce a false-negative result.

The antibody responses indicated in Figure 6 signify immunoglobulin G (IgG) antibody. Figure 7 depicts the immunoglobulin M (IgM) response as it may occur during HIV infection. Although IgM responses occur in most viral infections, no consistent IgM response seems to occur during HIV infection. Alternatively, it may be that IgM tests need improvement for better sensitivity. IgM assays, if effective in identifying infected individuals, would be beneficial for use during early infection (window period) before IgG antibodies are detected. In fact, some investigators have detected specific IgM anti-HIV in IgG-negative samples from individuals who have been later shown to seroconvert (i.e., early infection was detected). Because of the inconsistency of IgM assays, they have not been widely adopted. Currently, tests to detect IgM- and IgA-specific antibodies are being evaluated to determine the immune status of newborns suspected of being infected (see below).

Monitoring individuals at risk of acquiring HIV infection, both before and during infection, has provided some insight into the profiles of antibodies produced during the seroconversion process. In general, antibodies to the gag

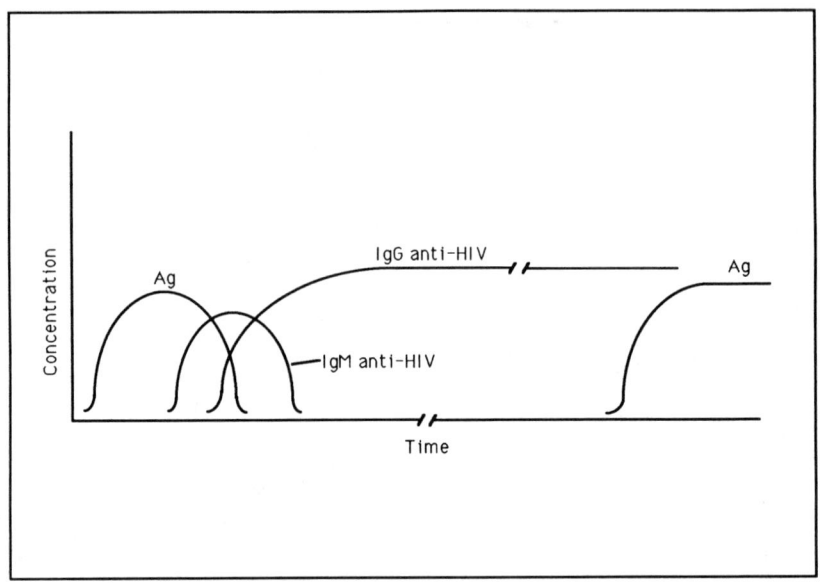

FIGURE 7. IgM response in HIV infection.

proteins, p55 and p24, are the first to appear via the use of most routine screening and confirmatory assays. In many individuals, only antibodies to p24 are produced initially. These profiles are most commonly noted in individuals who are in the early stages of HIV infection, and therefore individuals with these antibodies should be carefully monitored for evidence of future seroconversion. However, these patterns of antibody reactivity are also the most commonly observed patterns in individuals who are not infected. Indeed, a large number of normal, noninfected individuals will have antibodies that react with p24 and/or p55 antigens. The reason for the presence of this reactivity in noninfected individuals is unknown, but may represent cross-reacting antibodies directed against an unknown retrovirus or perhaps as a result of other diseases such as the autoimmune diseases. Sometimes in infected individuals, the first antibodies to appear are those against the env or pol antigens. This may be due to the strain of infecting HIV, or simply that certain individuals may respond more strongly and earlier to these antigens.

It is also important to consider that an individual with inconclusive patterns of reactivity may be infected with one of the other retroviruses that are known to cause human disease. Infection by HTLV-I/II (Chapter 6) may elicit antibodies to the core proteins that react with the core antigens of both viruses (i.e., cross-reacting antibodies). This is also true for infection by HIV-2; however, antibodies to the pol antigens are usually also present in HIV-2 infections (Chapter 5).

B. Newborns

HIV infection in infants born to infected mothers presents diagnostic problems because of the presence of maternally acquired HIV antibodies in the newborn's serum. IgG antibodies from an infected mother cross the placenta and will be detected when tests are performed on the blood of the newborn. Therefore, it is virtually impossible soon after birth to determine if a newborn is infected based solely on the presence of IgG antibodies. In most cases the antibody profiles of the mother and the newborn are identical. After approximately 8 to 14 months maternal antibody wanes in the infant, and a diagnosis becomes possible. In addition to the maternally derived serum antibodies, antibodies can be transferred postnatally through breast milk for many months after birth.

Presently, an infant born to an infected mother must be monitored over time using standard IgG antibody tests before the status concerning infection can be determined. Once maternal antibody has disappeared, the infant may serorevert (become negative after testing positive). However, it may be several more months before the newborn's immune system is competent enough to produce antibodies, and the newborn may still subsequently seroconvert, indicating true infection. Therefore, a newborn must be monitored for nearly 18 months before the absence of infection is verified. Approximately 20 to 30% of babies born to infected mothers will be truly infected with HIV. The reasons that most newborns of infected mothers are not infected with HIV are unknown.

Similar to that noted in adults, HIV infection in the newborn can be suggested, although not confirmed, if there is an increase over time in the number of different antiviral antibodies. This may be demonstrated by parallel testing of samples collected at different time intervals. Infection can only be confirmed at the time the antibody profiles in the newborn meet the criteria for positivity. Similarly, when maternal antibody is suspected in the newborn, demonstration of a rise in antibody titer over time in the newborn indicates infection.

The detection of specific HIV IgM antibodies that do not cross the placenta may be of value in diagnosing HIV infection in the newborn since this implies new antibody production in response to primary HIV infection. However, current assays for the detection of IgM antibodies have not been shown to produce consistent results, and therefore are not used routinely. In many seroconverting individuals and in newborns, IgM antibodies cannot be demonstrated. Therefore, the absence of specific IgM antibodies cannot be considered representative of noninfection. It may be that the tests currently available for detecting IgM antibodies are not adequately sensitive; they also may lack some specificity. It is not known exactly when specific IgM antibodies are produced in the infant, but it is thought to be around 6 months of age. The results of a positive IgM assay must also be viewed with caution, since interfering substances such as rheumatoid factor (RF) may produce false-positive results. RFs are IgM antibodies that react with IgGs. If these are present in the newborn's serum, they may bind to the mother's specific anti-HIV of the IgG class that have reacted in the test. This will interfere by producing reactive results when anti-IgM conjugates are used.

Recently, promising results have been obtained when testing the sera of infants using anti-HIV IgA-specific assays. The detection of IgA antibodies to HIV signals infection because these antibodies do not cross the placenta, and therefore cannot be of maternal origin. However, IgA antibodies to HIV may not appear very early, and the detection of IgA antibodies seems to be of more value when testing an infant at 3 months of age. In one large study, the sensitivity and specificity of the IgA assay for children older than 3 months were 97.6 and 99.7%, and the positive and negative predictive values were 99.4 and 98.7%, respectively (Chapter 7).

To perform anti-IgA or IgM HIV serological tests, serum must first be treated to remove IgG antibodies (of maternal origin, and which may interfere with the assay). By performing serial absorptions of IgG, and using a conjugate directed toward IgA or IgM, class-specific HIV antibody can be detected.

Newer, more specific, but technically difficult assays such as the polymerase chain reaction (PCR), are currently helpful in detecting the presence of the viral nucleic acid in the lymphocytes of newborns (Chapter 4). Viral culture and HIV antigen assays may also be of value for identifying HIV infection in newborns. Although sophisticated techniques such as PCR offer alternatives for the demonstration of HIV infection in newborns, they are not without limitation and must be performed by experienced laboratory personnel. Also, these assays are of little value in the neonate, but their sensitivity increases as the age of the infant advances beyond 6 months. In the absence of specific virological confirmation, monitoring of immunoglobulin levels, the CD4/CD8 ratio, and clinical signs can usually identify HIV infection in about 50% of infected children by 6 months. As in adult HIV infection, pediatric infection also produces a hypergammaglobulinemia. Furthermore, monitoring mothers for predictive markers such as CD4+ T cell levels and p24 antigenemia may indicate which mothers are most likely to deliver infected infants. In contrast to the long latency of HIV infection in adults, most cases of pediatric HIV infection progress to AIDS within 4 years.

II. CELLULAR CONSIDERATIONS DURING INFECTION

A. Lymphocytes, T Cell Subpopulations, Cell Interactions

A thorough understanding of the immune system is necessary to fully appreciate the way in which the virus infects and paralyzes the protection afforded by the immune system. The following brief description of the system concentrates on the major immune cells and the basic effects following HIV infection. The major components of the immune system include:

1. Macrophages, large cells that phagocytize, process and present antigens to other immune cells.

2. T lymphocytes, cells that play a vital role in cell-mediated immune reactions; consist of several subsets or subpopulations.
3. B lymphocytes, cells that are responsible for antibody production; usually require help from T lymphocytes.
4. Plasma cells, small cells that are produced from the maturation of B cells, and produce large quantities of antibodies.
5. Natural killer (NK) cells are active against tumors.
6. Polymorphonuclear leukocytes (PMN); phagocytic cells that are present in high numbers.

The T lymphocytes mature in the thymus and can be further classified into subpopulations known as CD4+ T cells (T4) and CD8+ T cells (T8). The CD4+ T lymphocyte subpopulation includes cells that can act either as helper or inducer cells. The CD8+ T lymphocyte subpopulation can function as either suppressor or cytotoxic cells.

CD4+ T helper (T_h) cells are responsible for providing assistance to other immune cells via direct contact, or by elaborating soluble substances (lymphokines/cytokines). This assistance to other cells can include events that bring about the maturation, activation, and proliferation of other T cells. In other words, CD4+ T_h cells can cause CD8+ T suppressor cells to suppress the activities of other cells (other T and B cells). They can cause CD8+ T cytotoxic cells to kill virus-infected cells. Similarly, CD4+ T_h cells cause B lymphocytes to mature to plasma cells and produce antibody; NK cells can be induced to kill tumor cells, and macrophages will become activated, resulting in increased phagocytosis, killing, and processing of antigen. The basic interactions of the cellular components of the immune system are depicted in Figure 8.

As an infectious agent such as HIV enters the body, it is usually first detected by the scavenger cells, the macrophages. The foreign agent is phagocytized by the macrophage, and the antigens of the agent are degraded and processed. Subsequently, these antigens are presented to the CD4+ T_h cells in such a way that the T cells recognize the antigens and respond. Macrophages also produce soluble substances that activate the CD4+ T_h cells. The T_h cells later elaborate their own mediators and these products influence the other immune cells, as mentioned above. The role of these soluble mediators in the immune response is now being realized as important, and assays have been developed to quantitate and evaluate each.

From this simplified description of the events in the immune system, it is apparent that when the CD4+ T_h cell is compromised by infection with HIV, the entire immune system is affected. Eventually, CD8+ T cytotoxic cells and the macrophages become incompetent due to the lack of support by T_h cells, and opportunistic infections are manifest; tumors may arise (possibly due to the decrease in NK activity), and antibodies to commonly encountered organisms may not be produced. In addition, immune cells may be destroyed as they fuse in masses (syncytia) composed of infected macrophages, CD4+ T cells, and other cells.

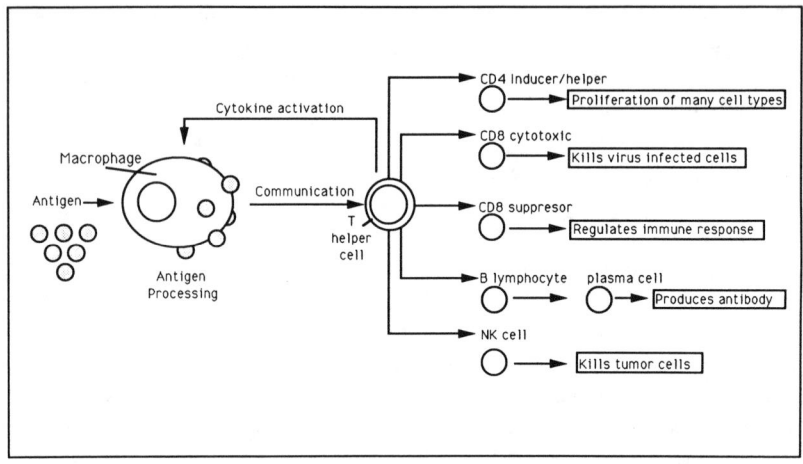

FIGURE 8. Basic interactions of the immune system.

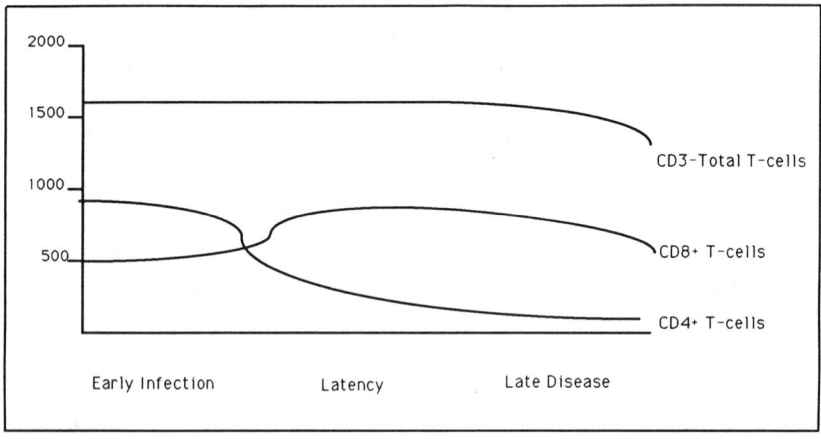

FIGURE 9. Relationship of T lymphocytes during HIV infection.

As HIV infection progresses, a general lymphopenia (decrease in lymphocytes in the blood) occurs. The primary effect is a decrease in the CD4+ T lymphocyte population (Figure 9). CD8+ T suppressor cells may initially increase and sometimes remain at high levels for long periods. The CD4+ T cell count offers the most information about the clinical disease stages, and therefore is the most widely used parameter to assess disease progression. The CD4+ T cell count is reduced in about half of the asymptomatic HIV infected individuals and in nearly all individuals suffering from ARC and AIDS.

Individuals with HIV infection and AIDS frequently possess several abnormalities in their immune system. This is not surprising because the virus infects cells and interferes with the normal interactions and regulation of immune cells. It is the CD4+ T lymphocyte population that is most dramatically affected, and consequently causes the severe immunodeficiency seen later in the infection. It is important to monitor infected individuals in order to determine their immune status and to gain some insight into the extent of involvement of the immune system. Often, the extent of involvement can be used as an indicator of disease progression. Severe changes can often be used as an adjunct to confirm HIV infection, if other reasons are ruled out.

Several different cell determinations can help to define cellular immune status. Most often, these include quantitations for CD4+ T cells, CD8+ T cells, and sometimes CD2+ and CD3+ cells. In addition, the relative proportion of CD4+ to CD8+ cells (the T4/T8 or helper/suppressor ratio) can be a very valuable measurement. If a monoclonal antibody against CD4 is used, all T cells possessing this marker (i.e., T_h and T inducer cells) will be identified. Likewise, CD8 monoclonals will identify T cytotoxic and T suppressor cells. Cells having the CD3 marker are all T cells considered to be immunocompetent (functioning properly). CD2+ cells represent the total number of T cells, regardless of their capability to function properly.

B. Cytokines

Cell-to-cell communication is an important part of the normal homeostatic mechanism of cell-mediated immunity (CMI). Control and coordination of CMI is achieved through the action of plurifunctional protein mediators known as cytokines. Cytokines are described in terms of their cellular origin, i.e., monokines (from monocytes) or lymphokines (from lymphocytes), or by their effects — interleukins, interferons, growth factors, chemotactic factors, colony stimulating factors, and tumor necrosis factor. In many HIV-infected individuals the normal balance of cytokines no longer exists. Even before the CD4+ T lymphocyte population decreases, these cells appear to function abnormally, particularly in the production and release of cytokines. The abnormal production and/or release of cytokines contributes to the immune deficiency characteristic of HIV infection and AIDS.

Cytokines are produced by many cell types, including macrophages and T lymphocytes. Their production follows many types of cellular injury, including that caused by infectious agents, and is the primary mechanism by which the CMI response is orchestrated. Immediately following any injury, inflammatory cells are recruited to the injured site by a process known as chemotaxis. Once the inflammatory cells are in the area, these cells then act to help eliminate the cause of injury, while growth factors from other cells begin the healing process. Once the injury is resolved, cytokine levels usually fall below detectable limits, as in healthy individuals.

The CD4+ T lymphocyte and the antigen-presenting macrophage play a central role in CMI. Cytokines produced by activated macrophages include tumor necrosis factor-α (TNF-α), interleukin-1β (IL-1β) and IL-8, as well as other monokines. These are usually produced within several hours following

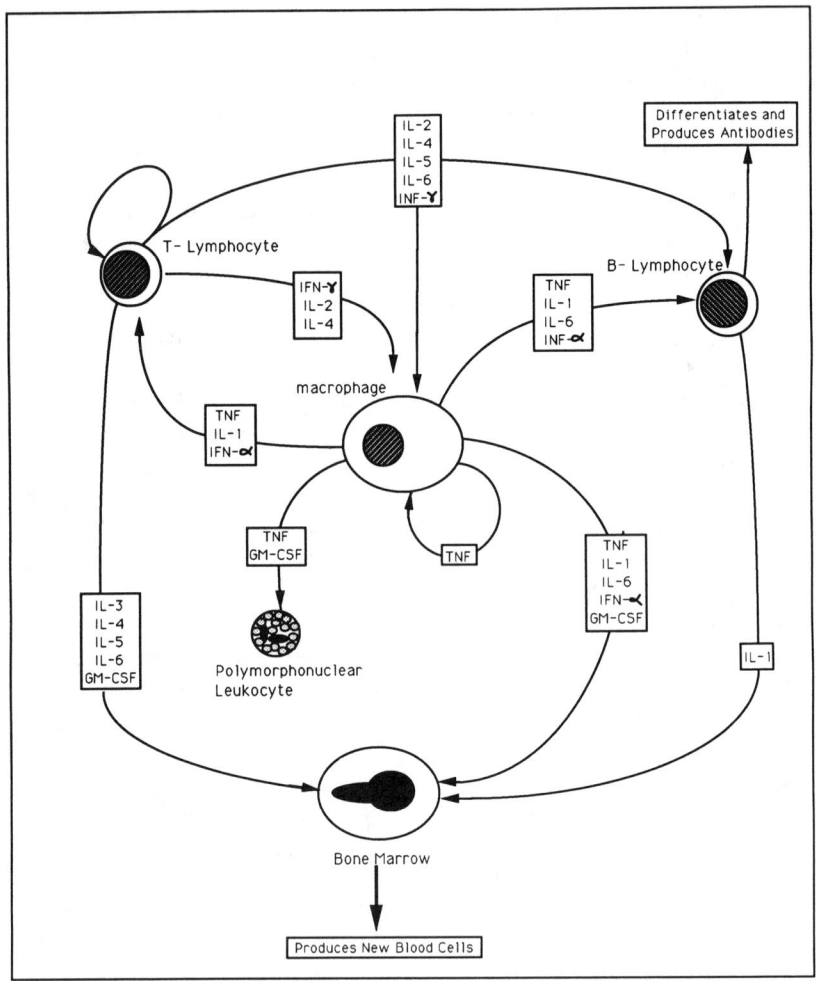

FIGURE 10. Cytokine actions.

interaction of the CD4 receptor with the virus. The activated CD4+ T lymphocyte can produce IL-1, as well as IL-2, -3, -4, -5, and -6, granulocyte-macrophage colony stimulating factor (GM-CSF), interferon-alpha, beta, and gamma (IFN-α, β, γ), as well as other lymphokines. Figure 10 indicates some of the cytokines and their interactions within the immune system.

Many nucleated cells have specific cytokine receptors for the various cytokines that are normally expressed on the outer surface membrane of the resting cell. The binding of a cytokine to its receptor is the first step in the initiation of cytokine action on a cell. An activated cell may then express additional surface receptors for several different cytokines, thereby causing the cell to be more receptive to further cytokine effects.

In some chronic inflammatory conditions, including HIV infection, the antigenic stimulus is not removed by the normal CMI mechanism. Continued immune activation by HIV and subsequent cytokine production creates a situation in which the target cells of the cytokines are continually stimulated. Increased expression of cytokine receptors can also be induced by continued activation; eventually, receptors can be released into the serum or body fluids in a soluble form; e.g., soluble IL-2 receptor (sIL-2r). Thus, detection of soluble cytokine receptors in serum can act as a serological marker of cellular activation.

TNF-α and IL-1 from the macrophages activate the CD4+ T cell but also induce viral replication in both infected T cells and macrophages. TNF levels have been shown to increase with the progression of HIV disease. The activated T cell in turn produces IL-1, -2, and -6, IFN-γ, and GM-CSF. These cytokines can activate both macrophages and other Th cells, induce an increase in receptor expression, and lead to even more viral replication in infected cells. The combination of increased cytokine activity and increased viral load can activate more macrophages with further production of monokines. Substantial evidence exists that TNF-α and HIV-1 can induce the expression of each other, resulting in a cytokine loop, the result of which is overproduction of cytokines. Elevated cytokine levels have been implicated in a number of disease processes including infection with HIV. Elevated levels of TNF, IL-1, -2, and -6, and sIL-2r, have all been reported in HIV-infected patients. Furthermore, reports have linked the levels of all of these at one time or another with disease stage and progression. Elevated levels of TNF have been implicated in the cachexia or "wasting syndrome" in AIDS patients as well as hematopoietic suppression that results in anemia, lymphopenia, and leukopenia. Figure 11 depicts a possible mechanism for the cytokine loop in HIV infection. How TNF actually increases HIV expression is unknown, but it is thought to occur through the action of a cellular protein known as NF-kB, which binds to the LTR portion of the viral genome.

C. Laboratory Investigation of Cellular Abnormalities
1. Cytokine Assays

The quantitation and evaluation of cytokine activity during different stages of HIV infection can lead to an increased understanding of the pathogenic mechanisms of the disease and may also help in evaluating the efficacy of new treatments or drug therapies. Quantitative ELISA techniques can be utilized for the determination of soluble levels of cytokines as well as their soluble receptors in serum, plasma, or in cell culture supernatants. The quantitation of membrane-bound cytokines or membrane-bound receptors for cytokines on activated cells is possible using flow cytometry (see below). A third method of detecting cytokine activity is by using the Northern blot (electrophoresis, blotting, molecular probes) technique to detect cytokine mRNA as it is produced in activated cells. At present, these assays are available for research use only and are not available for clinical diagnosis. Finally, there are functional

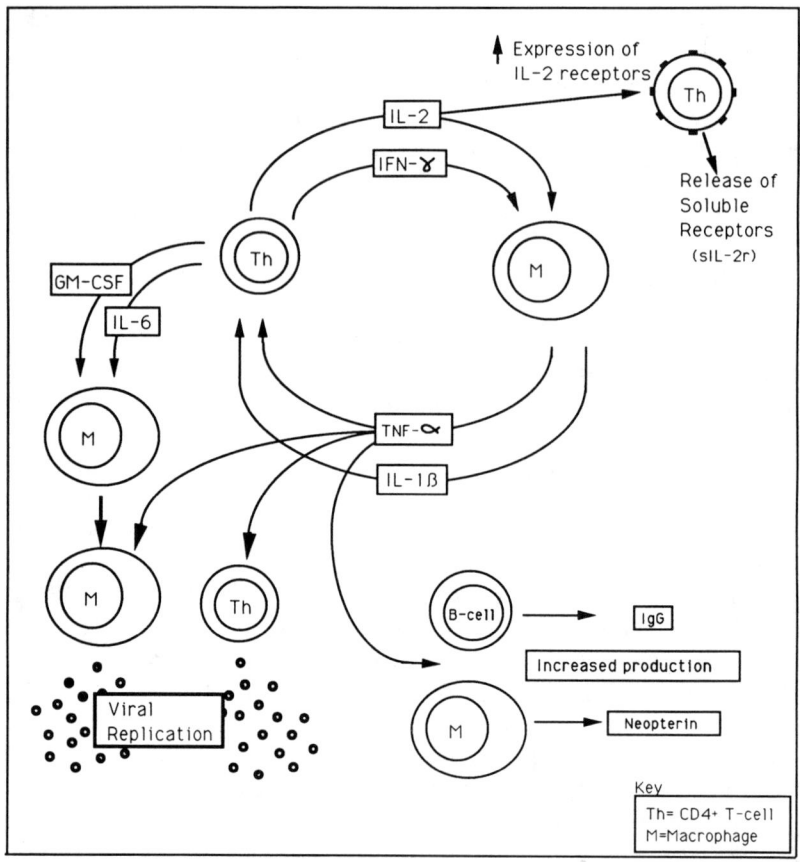

FIGURE 11. The cytokine loop.

assays to determine which cytokines are produced. At present, these types of assays are difficult to perform and are not addressed here.

2. Methods to Quantitate T and B Lymphocytes

The methods used to quantitate and evaluate immune cells are generally referred to as lymphocyte immunophenotyping methods, because the phenotype (physical identification) of the cells is determined. These methods are not easy to perform, and are time consuming and expensive.

In general, specific monoclonal antibodies made against a specific antigen present on the cells, such as the CD4 molecule, are labeled with a fluorochrome (a substance that fluoresces under UV light). The monoclonals are allowed to react with the mononuclear cells (lymphocytes and monocytes) from an individual and the cells that react can be classified into populations and subpopulations depending on which monoclonals are bound. The fluorescence of the labeled cells can then be detected. Some laboratories use a manual method that requires separation of the mononuclear cell population, addition of the monoclonal antibodies, followed by counting of the population by fluorescence microscopy. In this case, one aliquot of cells labeled with one monoclonal marker is counted first, followed by another aliquot using a different marker. The percentage of cells fluorescing in each aliquot determines the percent of each subset. This process of manual counting is a very time consuming and rigorous process, and may not be reproducible.

Alternatively, a sophisticated and expensive instrument called a flow cytometer can be used to determine percentages of the cell subpopulations by flow cytometric analysis. The flow cytometer is capable of automatically counting different subsets simultaneously as the cells flow through a detector. In principle, the basic method consists of injecting cells in suspension through a nozzle into a flowing sheath fluid that centers the cells in the stream. The cells pass single file through a laser light beam generated from a light source such as a mercury arc lamp. Each cell will generate an emitted signal from either the fluorescent marker or by light scatter. Appropriate filters and photomultiplier detectors transform the fluorescence or light scatter into electric signals that are subsequently converted by a computer into a digital readout. Figure 12 depicts the principle of the flow cytometer.

The flow cytometer can analyze two or three different color markers and two light scatter measurements simultaneously. The usual method is that one cell type is labeled with a fluorochrome of one color, while another is labeled with a different color. For example, the CD4+ T cells may be detected by using a fluorescein isothiocyanate (FITC; a green fluorochrome) -labeled monoclonal, while the CD8+ cells are simultaneously detected using the label phycoerythrin (PE), which produces a red color. The instrument can differentiate these fluorochromes and produce an accurate count of the various cell numbers. In addition to counting different cell populations, fluorescent markers can be used to determine which cell populations may be in a state of activation (activation markers). Some of the most important markers used to evaluate cells are listed in the next section describing normal values.

Cells can be identified not only by their markers but also by their size, granularity, and their cellular and cytoplasmic characteristics (by light scattering measurements). For example, large granular cells that contain the CD4+ marker can be differentiated from small, nongranular cells that also contain the CD4+ marker. Therefore, many different variables can be examined at the same time. Cells that contain the CD3+CD4+ markers would include the CD4+ T cells but would exclude the monocytes, which are CD3-CD4+. Similarly, CD8+CD3+ cells would include the CD8+ T cells but not the CD8+CD3- NK cells. The use of these dual markers to identify cell populations results in an improved means to evaluate specific cell populations. The flow cytometer is fully automated, but requires an operator thoroughly trained in its operation. The results obtained on the flow cytometer are much more accurate than the

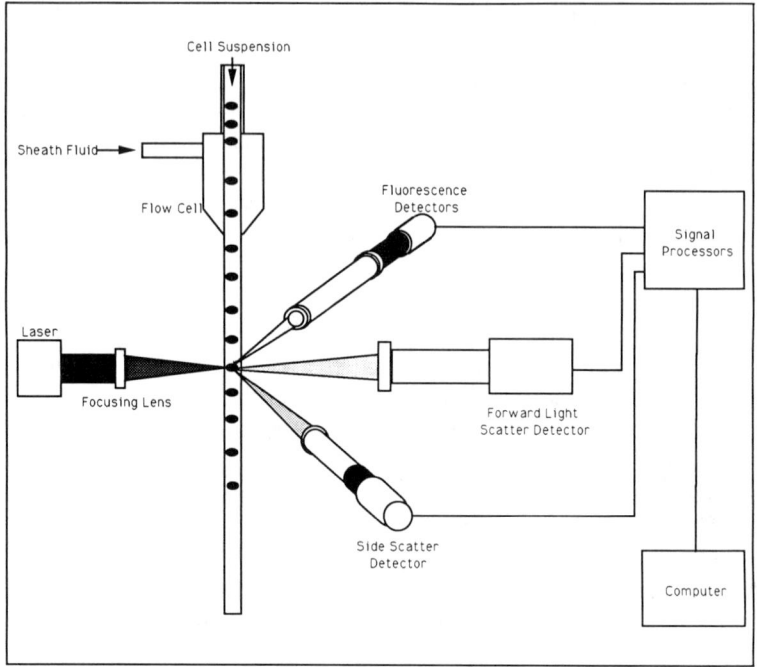

FIGURE 12. The flow cytometer.

manual method because many more cells can be counted objectively, thus reducing error. Furthermore, data can be stored electronically and analyzed at a later time.

Lymphocyte phenotyping provides valuable information concerning the immune status of individuals with HIV infection and AIDS. Unfortunately, the expense and the need for available experts to maintain and service a flow cytometer do not allow most laboratories to consider it for adoption.

3. Normal Values

Awareness of the normal values for these cell subpopulations makes it possible to determine whether the immune system has been affected by HIV infection. The following list indicates the normal adult values (percents and ranges of absolute values) for each of the subpopulations. Note that these numbers are only guidelines and that values may differ depending on the population tested, the laboratory, and the method used to quantitate. Normal values for these cells should be determined in each laboratory and should be representative of the population tested in age, sex, and race. Separate ranges should be determined for pediatric populations.

Cells	Marker	%	Absolute no.
Total T	CD2	74–92	752–2533
Immunocompetent T	CD3	67–84	591–2452
T_h/inducer	CD4	39–58	408–1444
T–suppressor/cytotoxic	CD8	14–36	163–860

To calculate the absolute values:

1. First obtain the total white blood count (WBC), and the percent of lymphocytes of the patient. These are usually obtained by using a Coulter® counter, or manually using a hemocytometer.
2. Multiply the number of WBC by the percentage of lymphocytes. This equals the total number of lymphocytes.
3. Multiply the total number of lymphocytes by the percent of each subset as determined by the flow cytometer. This yields the absolute numbers of cells in each subset.

For example:

WBC count = 3900/µl
% Lymphocytes = 35%
Total lymphocytes = 3900 × 0.35 = 1715/µl
If CD4+ cells = 18%
Then, the absolute number of CD4 cells = 1715 × 0.18 = 309 (there are 309 CD4+ T cells)

Other important markers that may be used in flow cytometric analysis to evaluate the immune system include:

CD14	Monocytes
CD16, CD56, CD57	NK cells
CD19, CD20	B cells
CD45	All leukocytes
CD25 (IL-2R)	Activated T cells
CD38	Marker of activation and immaturity

Evaluation of the immune system using these markers requires a great deal of knowledge and expertise. Many of the cells of the immune system may contain the same marker even though they represent different cell populations. For example, some T lymphocytes also contain the CD56 marker that is common to NK cells. Therefore, when evaluating cells using this marker, T and

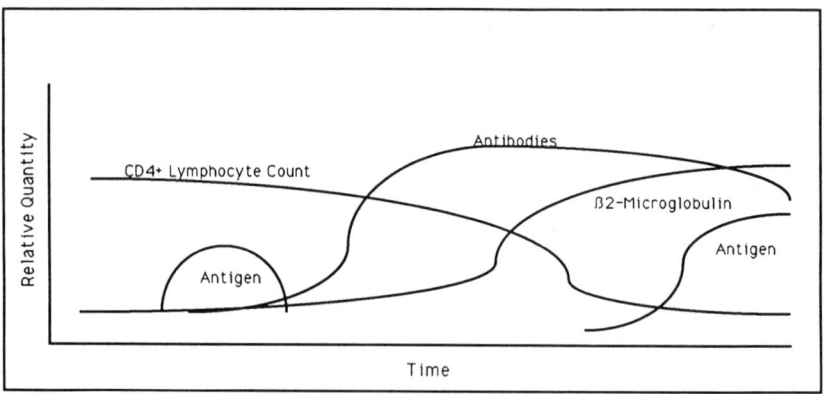

FIGURE 13. Relationship of B_2m, CD4+ T cells, antigen levels, and antibody responses during HIV infection.

NK cells can be enumerated simultaneously using dual markers to ensure that the T cells that are CD56+ are not included in the NK cell count and in the T cell count. The system is complex. Quality control measures must also be included in these evaluations. For example, an isotype control must be used to control for nonspecific binding of the monoclonals to the patient's cells; the enumerations can be checked by determining the "lymphosum" (i.e., the sum of T + B + NK should approach 100%). Because of the variety of instruments and combinations of monoclonal markers available, it is not possible to describe the details of the necessary quality control measures required for cell phenotyping. Guidelines are available from organizations such as the National Committee for Clinical Laboratory Standards (NCCLS), the CDC, and the Association of State and Territorial Public Health Laboratory Directors (ASTPHLD).

4. Expected Values during HIV Infection and AIDS

Evaluation of different cell populations such as lymphocyte subpopulations provides a valuable assessment of immunologic damage and can be used to monitor patients with HIV infection and AIDS. Since the course of disease is highly variable, many authorities recommend that the CD4+ T cell population be monitored at 3 to 6 month intervals following detection by infection. CD4 and CD8 markers are the most commonly evaluated, although the CD38 marker was recently shown to provide important information for predicting survival in AIDS patients.

Normally, T lymphocytes (CD4+ and CD8+) account for 70 to 90% of all lymphocytes. Their numbers are generally near 1600 cells per cubic millimeter (μl), with CD4+ cells accounting for about 1000 and CD8+ about 500. The decrease in CD4+ T cells noted during HIV infection is usually gradual (50 to 100 cells per cubic millimeter/year), with levels being about 600 within the first year after infection (Figure 13). CD4+ T levels are highly prognostic for predicting survival, especially during late disease. Approximately 80% of individuals will die within 1 year when the CD4+ T cells fall to 10%. High levels of CD38+CD8+ cells are characteristic of late HIV disease and are highly prognostic for HIV disease progression.

Although no consensus can be reached on the exact value that indicates disease, most authorities agree that a CD4+ T cell count of <500 in an HIV infected individual is highly suggestive of immunodeficiency, if other reasons for the decrease have been ruled out. Generally, the lower the CD4+ T cell value, the more severe the disease progression and the less favorable the prognosis. It is often helpful to determine the T4/T8 ratio. Since CD4+ cells account for about 60% of the T lymphocytes, and CD8+ cells account for about 30% of the T lymphocytes, a normal ratio is always >1.0 and usually >1.5. Therefore, there is a significant decrease in CD4+ T lymphocytes when the ratio falls below 1.0 (a reversal of the ratio).

In summary, typical immunophenotypic profiles are produced during HIV infection and AIDS, and the changes are progressive. Characteristically there is lymphopenia, a low CD4+ T cell count, a low CD4+/CD8+ T cell ratio (reversal), and a high proportion of CD38+CD8+ cells. Increased percentages of the CD38+CD8+ cells seem to identify those individuals who will have a more rapid and aggressive disease.

D. Functional Defects of Immune Cells

Cells can be present in sufficient quantities, yet may not function properly (functional defects). In many disease states immune cells do not perform their required activities. Functional defects in cells are present in HIV infection and particularly in AIDS. Since functional assays are not routinely performed in clinical laboratories, they are mentioned only briefly.

In some instances, quantifying parameters such as membrane antigen expression can provide an indication related to the functional activity of a cell population. For example, demonstration of the functional activation of cells can be accomplished by using the flow cytometer to quantitate the IL-2/CD25 receptors on activated T lymphocytes. Mononuclear cells such as T lymphocytes can be functionally characterized by determining their ability to undergo blast transformation following exposure in cell culture to a mitogen such as phytohemagglutinin (PHA), or *in vivo* by determining the T cell response to soluble antigens such as tetanus toxoid. Many of these functional defects are evident early in HIV infection. The response to PHA is a proliferation of the cells with incorporation of precursor nucleic acids that are needed for cell division. Measurements of blast transformation are usually accomplished by assaying for the incorporation of radiolabeled thymidine and by determining a stimulation index. Alternately, proliferation can be determined by flow cytometry and measuring the presence of IL-2 receptors. These transformation assays can be conducted in a similar fashion on monocytes and B lymphocytes.

Evaluation of B cells and B cell function does not generally provide much information of prognostic value. In HIV-infected individuals an increased number of B cells spontaneously secretes immunoglobulin (polyclonal activation). In addition, B cells from infected persons exhibit abnormal responses to activation signals (such as mitogens).

Monocytes and macrophages also exhibit impaired function during HIV infection. Among the defects are decreased chemotaxis, monocyte-dependent T cell proliferation, Fc receptor function, and C-3 receptor mediated particle clearance. Similarly, PMNs exhibit impaired function in HIV infection, including reductions in chemotaxis, bacterial killing, and phagocytosis.

E. Other Markers for Cell-Mediated Immunity during Infection

Various markers other than those already mentioned have been used to help evaluate the immune system during HIV infection. Some have shown promise when used for prognostic purposes. The most common markers used to help assess disease progression in infected individuals are beta-2 microglobulin (B_2m), neopterin levels, and p24 antigenemia. p24 antigenemia is a marker of viral activity and is discussed in Chapter 4.

B_2m is a soluble product, elaborated following immune (lymphoid) activation, which can be measured in serum. This polypeptide product is elevated in HIV infection and is strongly associated with cellular destruction and the risk of progression to AIDS. Levels of B_2m are measured by radioimmunoassay (RIA) or by ELISA. Figure 13 shows the relationship of B_2m levels to CD4+ T cells, antigen levels, and the antibody responses to HIV.

Neopterin is a product of macrophages that is produced when they are stimulated by IFN-γ from activated T cells. Neopterin levels are also elevated in HIV-infected individuals and correlate well with prognosis. Elevated levels (generally above 10 nmol/l) are generally independent of changes in T cell levels and peak during antigenemia in viral infections. The combination of CD4+ T cell levels and neopterin measurements is a better indicator of prognosis than either alone. Neopterin levels are measured by RIA, and can be determined in serum or urine.

REFERENCES

Landay, A., Ohlsson-Wilhelm, B., and Giorgi, J. V., Application of flow cytometry to the study of HIV infection, *AIDS*, 4, 479, 1990.

Laurence, J., Immunology of HIV infection. I. Biology of the interferons, *AIDS Res. Human Retroviruses*, 6, 1149, 1990.

Le, J. and Vilcek, J., Biology of disease, tumor necrosis factor and interleukin 1: cytokines with multiple overlapping biologic activities, *Lab. Invest.*, 56, 234, 1987.

Mandell, G. L., Douglas, R. G., and Bennett, J. E., *Principles and Practice of Infectious Diseases*, 3rd ed., Churchill Livingstone, Edinburgh, 1990.

Maury, C. P. J. and Lahdevirta, J., Correlation of serum cytokine levels with hematological abnormalities in human immunodeficiency virus infection, *J. Intern. Med.*, 227, 253, 1990.

Merrill, J. E., Koyonagi, Y., and Chen, I. S. Y., Interleukin-1 and tumor necrosis factor-alpha can be induced from mononuclear phagocytes by human immunodeficiency virus type-1 binding to the CD4 receptor, *J. Virol.*, 63, 4404, 1989.

Merrill, J. E. and Chen, I. S. Y., HIV-1, macrophages, glial cells, and cytokines in AIDS nervous system disease, *FASEB J.,* 5, 2391, 1991.

Quinn, T. C., Kline, R. L., Halsey, N., Hutton, N., Ruff, A., Butz, A., Boulos, R., and Modlin, J. F., Early diagnosis of perinatal HIV infection by detection of viral-specific IgA antibodies, *JAMA*, 266, 3439, 1991.

Rautonen, J., Rautonen, N., Martin, N. L., Phillip, R., and Wara, D. W., Serum interleukin-6 concentrations are elevated and associated with tumor necrosis factor-α and immunoglobulin G and A concentrations in children with HIV infection, *AIDS*, 5, 1319, 1991.

Rosenberg, Z. F. and Fauci, A. S., The immunopathogenesis of HIV infection, in *The Human Retroviruses*, Gallo, R. C., Ed., Academic Press, New York, 1991, 141.

Stites, D. P. and Terr, A. I., *Basic and Clinical Immunology*, 7th ed., Appleton & Lange, New York, 1991.

Wright, S. C., Jewett, A., Mitsuyasu, R., and Bonavida, B., Spontaneous cytotoxicity and tumor necrosis factor production by peripheral blood monocytes from AIDS patients, *J. Immunol.*, 141, 99, 1988.

Zangerle, R., Fuchs, D., Reibnegger, G., Fritsch, P., and Wachter, H., Markers for disease progression in intravenous drug users infected with HIV-1, *AIDS*, 5, 985, 1991.

Chapter 3

SCREENING TESTS FOR HIV-1 INFECTION

I. INTRODUCTION

No available serological tests will diagnose AIDS, only HIV infection. Thus, these tests should never be referred to as "AIDS tests". Anti-HIV screening assays were primarily developed in an effort to protect the blood supply. Testing has since been required in some situations for obtaining life insurance, marriage licenses, immigration documents, and for seroprevalence surveys. Testing of individuals for diagnostic purposes has three major applications: (1) to verify that a patient presenting with symptoms compatible with HIV infection truly is infected; (2) to monitor a patient who has been exposed to HIV for evidence of infection; and (3) to test healthy individuals who belong to a group at risk for infection. In addition, the recent introduction of treatments both before and after disease manifestations has favored the policy of early identification of all HIV-infected individuals. For all of these reasons, and the means by which HIV is transmitted, an enormous number of tests are performed throughout the world.

Serological screening tests designed to detect HIV antibodies are many in number and variety, and are the most common approach for detecting infection. The initial tests, developed in 1985, were ELISAs. ELISAs were chosen because these types of tests had been used successfully for a variety of other infectious agents. Also, ELISA tests are generally easy to perform, adaptable for testing large numbers of samples, do not require the use of radioactive substances, and are sensitive and specific. Presently, a large number of different commercial companies produce these tests in a variety of different formats. Although the tests that were initially developed were very effective in detecting HIV antibodies in patients with AIDS, their use for detecting antibodies in nondiseased individuals has presented a different challenge. The occurrence of even a small number of false results by these tests can have profound implications when testing a population at low risk for infection (see Section VI, "Predictive Values", of Chapter 7). This is especially true when testing blood donors, since false results waste resources in discarded blood units and require verification of reactive results using more expensive tests. In addition, a false-negative result indicating that an infected individual is not infected can have serious consequences for the recipient of the blood.

Not long after the HIV ELISAs were marketed, other techniques became available. Latex, red cell, and gelatin particle agglutination tests were introduced in order to offer alternatives for ease of performance and for cost savings. Subsequently, the simple-to-perform dot-blot assays were introduced.

Most of the tests that were initially produced and many of those still used are based on viral lysate antigens; the antigens used in the test are prepared from whole disrupted HIV virions. The lysates usually contain components of the host cells in which the virions are propagated. These cellular components in the assay system may lead to biologic false-positive results. Most commonly, these contaminants are derived from nonviral antigens, such as those of the major histocompatibility locus originating from the lymphocytes in which the virus is grown. Therefore, persons who have HLA antibodies as a result of exposure to fetal white blood cells during pregnancy or by blood transfusion could test reactive by the screening assays, but may not be infected by HIV.

Newer tests incorporate the use of recombinant or synthetic peptide antigens, and offer improvements in test performance. As technology evolved, better diagnostic tests have become available. The terms first, second, and third generation have been applied to anti-HIV assays depending on the source of the antigen that is used or on the format of the assay. First generation assays incorporate native viral antigens derived from the detergent disruption of viruses (viral lysates) that are grown in human lymphocytes; second generation assays use artifically derived recombinant antigens that are expressed from bacteria or fungi or chemically synthesized oligopeptides of about 15 to 40 amino acids (synthetic peptides). A detailed discussion of recombinant and synthetic peptide derived antigens is presented later in this chapter. Third generation tests have recently been introduced and generally incorporate synthetic peptide antigens in an antigen sandwich format (discussed below). In addition, combination assays are sometimes considered to be third generation tests, depending on the antigen used and the assay format.

At present, a variety of assays are available, and the choice of a test can be tailored to be appropriate for almost any given testing situation. Some can be performed in 5 min, while some require 22 h; many need sophisticated instrumentation while others need none; some require large volumes of serum (800 µl), some need only 1 µl; indicator systems range from color production to no color production for a positive test; others are read visually, some require a fluorescence detector or a spectrophotometer; some tests require a great deal of expertise to perform while others are so simple that a laboratorian with only the most fundamental skills can perform them accurately. The cost of these tests also varies from <$1 (U.S.)/test, to >$12/test.

Screening assays are designed to detect all positive sera (i.e., to be sensitive), even if some false-positives occur. The sensitivity of these screening assays has been optimized in order to identify all individuals who may be infected. The results are not to be used as the final interpretation, as sera from some individuals will inevitably produce reactive results, even though the individuals are not infected. This concept is further discussed in later sections.

During any testing procedure, mistakes may occur due to technical error; therefore, repeat testing is mandatory. A result must be repeatedly reactive at least two or three times before the sample is considered to be truly reactive by the screening assay. Sera that produce a positive result by screening assays are termed reactive, and if reactive by confirmatory tests, they are usually classified as positive.

Most of the screening tests that utilize a conjugate system incorporate an anti-human immunoglobulin as a means to detect the specific HIV antibody. This is usually directed against the IgG class of human immunoglobulin, since IgG is the major antibody that can be detected during infection. Recently, investigations have revealed that conjugates to detect IgM and IgA can be of value for demonstrating early infection and infection in the newborn, respectively. In fact, in one study performed recently, anti-HIV IgM antibody was the only antibody present in several samples from a group of individuals who had recently seroconverted. Therefore, the inclusion of anti-IgM and anti-IgA reactive conjugates may have the potential to improve the sensitivity of the assays for detecting early infection. Generally, commercial companies have not used these polyvalent conjugates, most likely because the production of IgM following infection is a transient response, and the binding of IgM may tend to be nonspecific, thereby interfering with the specificity of the assay. Also, the inclusion of these conjugates would require reoptimization of the assays and perhaps delays in their introduction into the market. However, third generation antigen sandwich assays may be of value in helping to address this concept, as discussed below. Furthermore, the exact kinetics and significance of IgM and IgA responses have not been systemically characterized.

Screening tests to detect antibodies to HIV-1 will not always detect individuals who are infected with HIV-2 (Chapters 4 and 5). The use of combination assays to detect antibodies to HIV-1 and HIV-2 simultaneously, and to differentiate infection by these two viruses has gained much popularity in many parts of the world. As the prevalence of infection by HIV-2 increases throughout the world, the need for these screening combination tests will become more obvious. A detailed description of combination tests is presented in Chapter 4.

II. PRINCIPLES OF THE TESTS

A. ELISAs

A variety of different ELISAs are available, but most of those used for detection of HIV antibodies are classified as either indirect, competitive, sandwich, or capture assays. Most of these types consist of HIV antigen attached on a solid phase (support), and incorporate a conjugate and substrate detection system. Solid supports are supplied by the kit manufacturer with the viral antigen (viral lysate, recombinant or synthetic peptide) already attached.

These solid supports can be "wells" of a microtiter plate, plastic beads, or a type of paper, usually nitrocellulose. Recently, microbeads having an increased surface area have been employed as the solid support in order to absorb larger quantities of antigen. Tests using these microbeads are available in dot-blot (Genetic Systems Genie) and semiautomated systems (Abbott IMx).

In the most popular type of ELISA (indirect), addition of sample to the antigen-coated beads or wells will bring about the binding of specific antibody (anti-HIV), if present. Following a wash step to remove unbound serum constituents, conjugate is added and incubated. Conjugate will bind to the patient's antibody (if present), with excess conjugate being removed during another wash step.

Conjugates are most often antibodies coupled to enzymes, fluorochromes, or other reagents that will subsequently bring about a reaction that can be visualized. The antibody portion of the conjugate is usually an anti-human immunoglobulin directed against either a human antibody (in the indirect assays) or directed toward the HIV antigen on the solid support (in the competitive assays). In either case, these conjugated antibodies will bind in the test system and result in the generation of a signal. In the case of enzyme conjugates, the signal subsequently generated will be a color reaction; with fluorochromes, the signal generated is fluorescence.

Enzymes used in enzyme immunoassays (EIA) such as the ELISA are usually either alkaline phosphatase or horseradish peroxidase. Although several substrates can react with these enzymes, 4-nitrophenylphosphate is the usual substrate for alkaline phosphatase, while *o*-phenylindiamine-2 HCl (OPD) or 4-chloronaphthol (4CN) is used for horseradish peroxidase. These enzymes are capable of modifying a substrate in the presence of a chromogen (color-producing compound) to produce a colored product that can be detected either visually or by an instrument (spectrophotometer). Therefore, in an indirect assay (see below), the addition of substrate will produce a color reaction if the conjugate has been bound to the patient's antibody, which is bound to the antigen on the solid support.

The colored product that is detected by a spectrophotometer is measured at a particular light wavelength at which the color absorbs the incident light. This results in a signal usually converted by the instrument into optical density (O.D.) units. O.D. readings are produced and related to the light transmitted through the colored product, and are a measure of the intensity of the colored product. Therefore, the amount of color produced by the enzyme-substrate reaction is determined by the readings in O.D. units.

Recently, Wellcome introduced a substrate amplification system designed to increase the sensitivity of the ELISA. In this system a cyclic enzyme amplification step increases the color reaction; i.e., the signal is amplified. Therefore, more color is produced for each molecule of antibody present. Wellcome states that the sensitivity of assays utilizing this amplification step is tenfold greater than that of the standard enzyme-substrate systems. The basic principle of this amplification step is presented in Figure 14.

All ELISA tests are easy to perform, but require careful adherence to procedures. Any deviation in incubation times, temperatures, or volumes can result in dramatic changes in test results. Wash procedures must be thorough, and reagents must be prepared exactly as indicated by the manufacturer. Conjugates must be thoroughly, but gently, mixed in order not to denature the enzyme. Substrates must be prepared immediately before use and should be kept in the dark prior to addition in the test system.

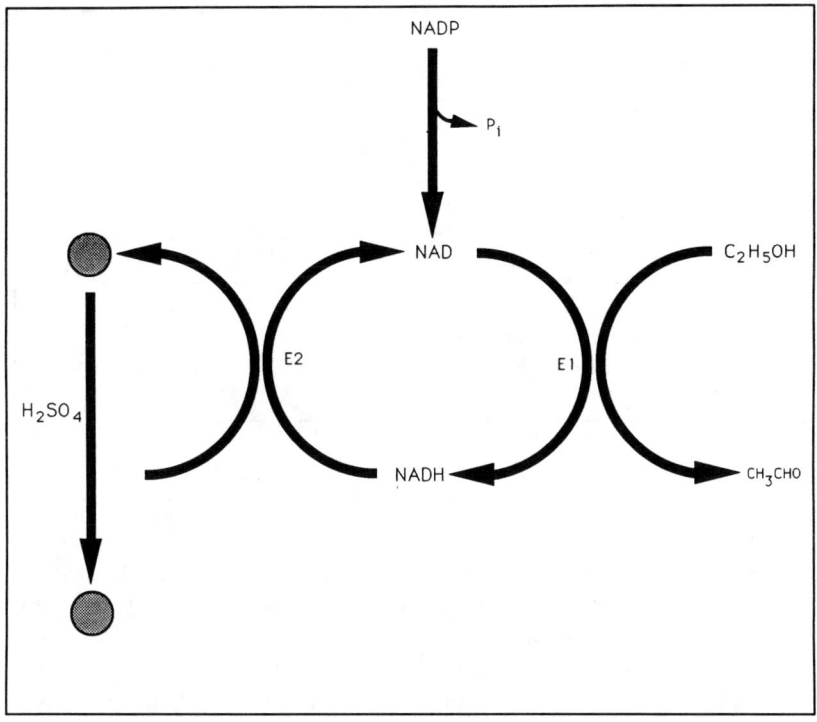

FIGURE 14. Signal amplification.

1. Indirect

In the indirect ELISA (or antiglobulin ELISA), test serum is added to the solid phase containing the antigen and is incubated for a specified time and at a particular temperature. If anti-HIV is present in the serum it will bind to the antigen on the solid phase. Following a wash step, conjugate (anti-human immunoglobulin labeled with an enzyme) is added and incubated. During this step, the anti-HIV antibody attached to the bound antigen will bind the conjugate. Another wash step removes excess conjugate, and substrate is added. During the subsequent incubation, the conjugate will modify the substrate to produce a color. The indirect ELISAs produce more color as the unknown antibody concentration in the sample increases (Figure 15). Conversely, with small amounts or an absence of antibody, smaller quantities of conjugate will bind, resulting in less substrate cleaved, less color produced, and a lower O.D. value. Therefore, the O.D. value is directly proportional to antibody concentration in the indirect ELISAs.

2. Competitive

Competitive ELISAs differ in that the antibody to HIV in the sample competes with the conjugate (which is an antibody also directed against the

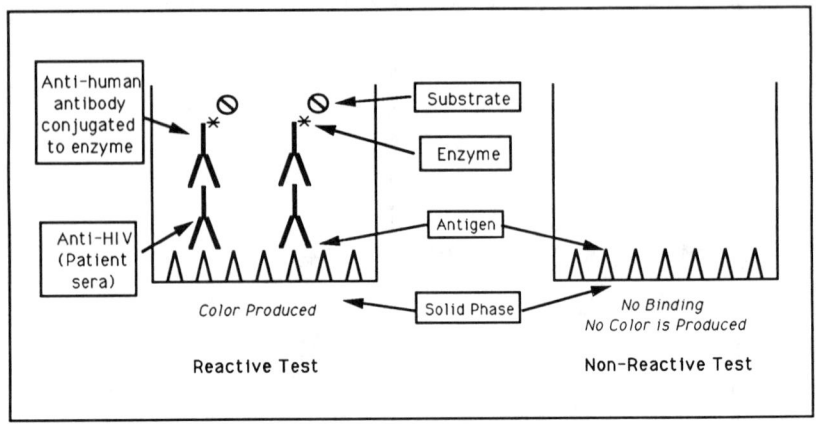

FIGURE 15. Principle of the indirect ELISA.

HIV antigen) for reactive sites on the bound antigen (Figure 16). Therefore, in competitive ELISAs, both the sample containing the antibody to HIV and the conjugate are added to the solid phase (containing the HIV antigen) at the same time. If the concentration of antibody in the sample is high, very little conjugated antibody can bind to the immobilized antigen because the two antibodies compete for the same sites on the antigen. Less color development will occur because conjugate has not bound and therefore is not available to modify the substrate. Conversely, with samples that contain little or no HIV antibody, more conjugate will bind to the antigen on the solid phase and the subsequent addition of substrate will cause more color development. Hence, the amount of unknown antibody in the sample is inversely proportional to the amount of color produced and the O.D. value.

Competitive ELISAs require less total time to perform since there are fewer steps (conjugate and sample are added together), and usually no predilution of the sample is required. Many persons consider competitive assays to be more specific than the indirect ELISAs; however, this point of view recently has been challenged.

3. Antigen Sandwich

Recently, the format of the classical ELISAs has been modified in an effort to gain sensitivity and specificity. Several types of modifications exist, but only the antigen sandwich type is presented here. Antigen, which is attached to the solid phase, binds antibody in the test sample. Since antibody molecules are bivalent they are still able to bind to another molecule. Therefore, in the next step a similar enzyme-labeled antigen is added and will attach to the same antibody molecule that is bound to the solid phase antigen. This forms a sandwich of antigen/antibody/labeled-antigen complex. The following step — addition of substrate with subsequent color development — is similar to that in other ELISAs. One major advantage of the antigen sandwich technique is

Screening Tests for HIV-1 Infection

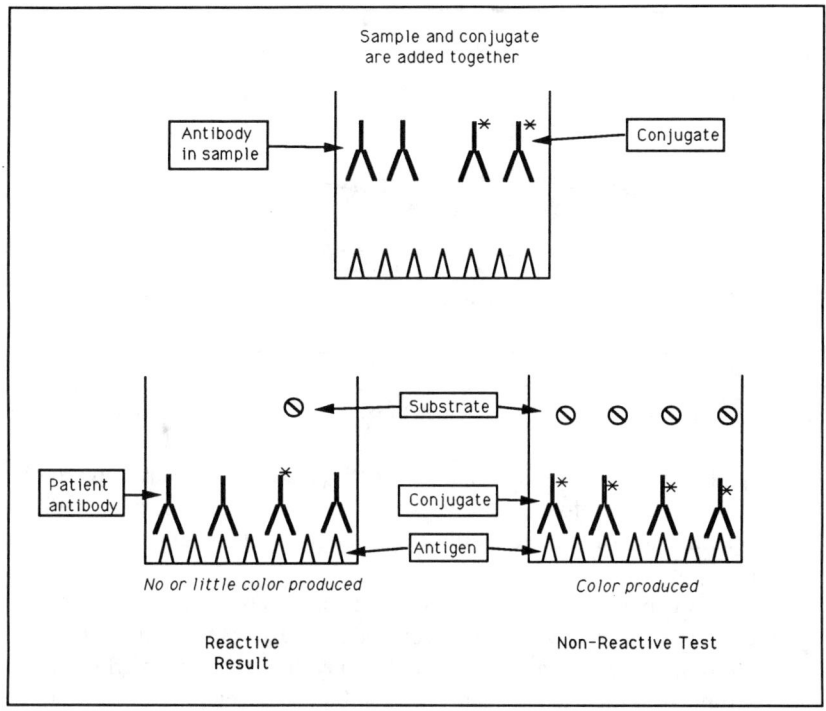

FIGURE 16. Principle of the competitive ELISA.

that all classes of antibody are detected, since there is no specific anti-immunoglobulin conjugate used. Wellcome as well as Abbott market this type of third generation assay for detecting antibodies to HIV-1 and HIV-2 (see Figure 36, Chapter 5). Other types of sandwich techniques such as antibody sandwich ELISAs have been developed for detecting p24 antigen (see Figure 31, Chapter 4). The principle of the antigen sandwich (or immunometric) ELISA for detecting p24 antibody is depicted in Figure 17.

4. Antigen and Antibody Capture

Antigen capture ELISAs can be of the indirect or competitive type, and differ only in the initial step of attaching antigen to the solid phase. A monoclonal antibody (very specific antibody directed toward only one antigenic determinant) is bound to the solid support. A viral lysate is added and the specific antigen(s) in the lysate that the monoclonal is directed toward is bound (captured) by the monoclonal. Therefore, the serum antibody detected by the captured antigens should be specific for the antigens and not directed toward unimportant antigens or contaminants in the lysate. This tends to decrease the amount of contaminating substances bound to the solid support, and hence decreases the amount of nonspecific binding. Antigen capture assays are considered to be more specific than the indirect assays, which incorporate total

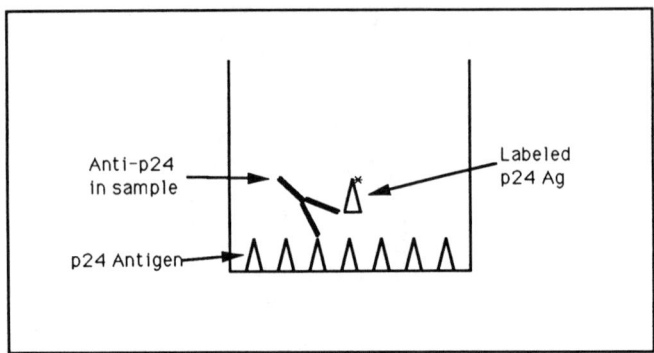

FIGURE 17. An antigen sandwich ELISA for detecting HIV antibody.

viral lysates. These assays are obtained with the antigen already captured by the monoclonal. Therefore, the procedure is performed identical to the indirect or competitive assays. Figure 18 illustrates the principle of an indirect antigen capture assay.

Antibody capture ELISAs (or class-specific antibody capture ELISAs) were recently introduced and were also developed in an effort to increase sensitivity and specificity. These assays were also designed for use in testing fluids other than serum, in which the total immunoglobulin concentration is low (e.g., saliva and urine). They are based on the capture of total IgG by incorporating an anti-IgG attached to the solid phase. Therefore, all antibodies of the IgG isotype, including specific anti-HIV, are bound. In essence, this is a means to concentrate the immunoglobulin before assaying for specific anti-HIV. Subsequently, a labeled HIV antigen is added that will bind to the anti-HIV IgG, if present. Addition of substrate will result in color production. The principle of the IgG antibody capture ELISA (GACELISA) is shown in Figure 19, and is discussed again later in this chapter.

B. Agglutination Tests

Tests employing agglutination as the indicator system have been used for diagnosing infectious diseases for many years because they generally have good sensitivity for detecting antibody. However, specificity is sometimes compromised. Agglutination tests can incorporate a variety of antigen-coated carriers. Carriers are particles used to support or "carry" the antigen.

HIV agglutination-based tests incorporate red blood cells (RBCs), latex particles, gelatin particles, autologous RBCs, or microbeads as carriers. HIV antigens (viral lysate or specific recombinant or synthetic peptides) are adsorbed onto the carrier, and these antigen-coated reagents come ready to use.

When antigens are bound without specific attachment (i.e., they are passively adsorbed onto the carriers) the technique is referred to as passive agglutination. If RBCs are used as the carrier system, the assay is called a passive hemagglutination assay.

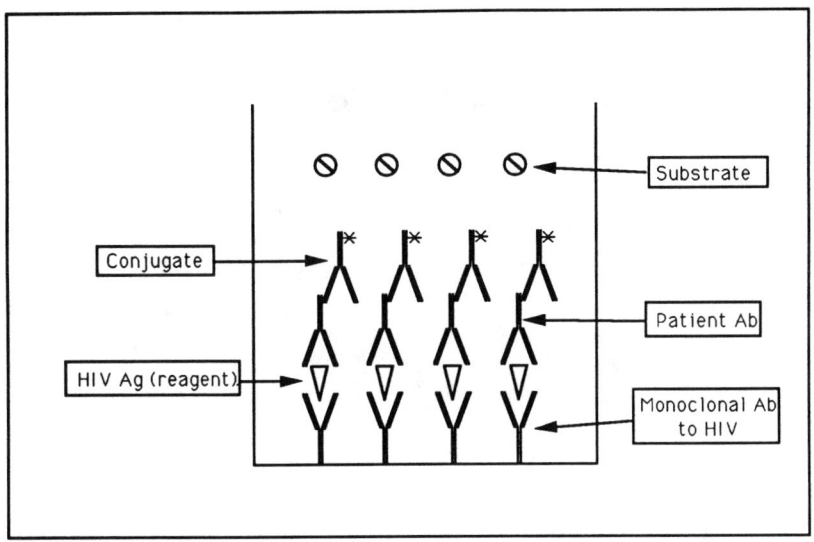

FIGURE 18. Principle of the indirect antigen capture ELISA.

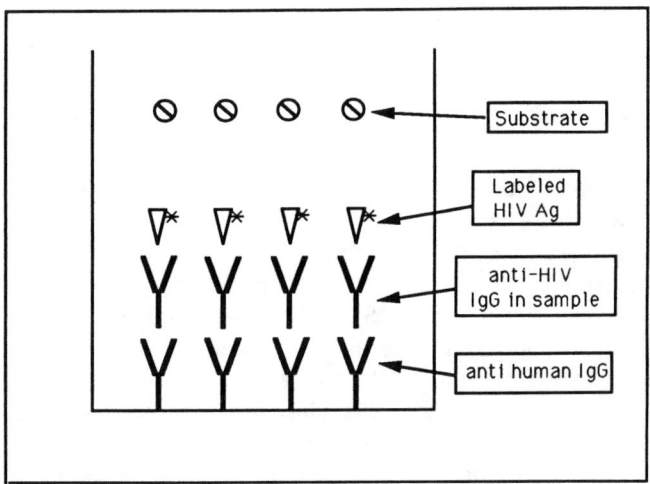

FIGURE 19. The IgG antibody capture ELISA.

During the agglutination reaction, a lattice network is formed between the antigen-coated particles and the antibody, as antibody in the sample reacts in the system. This reaction brings about the clumping (agglutination) of particles (Figure 20). The agglutination is macroscopic and is read visually, except in the microparticle assays.

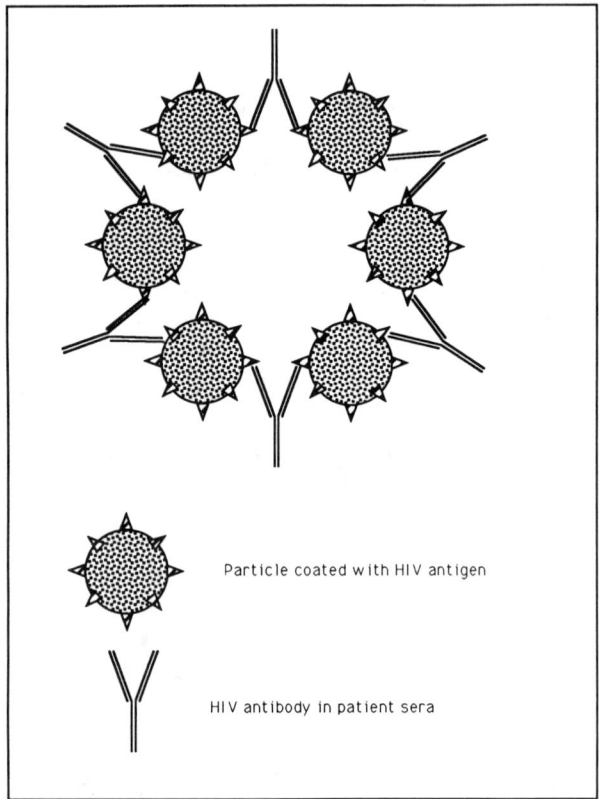

FIGURE 20. Formation of the lattice network and agglutination.

One possible problem with agglutination assays is the phenomenon of prozone reactions. Prozone refers to the inhibition of agglutination when excess antibody is present, thereby preventing the optimal combination of antibody and antigen. In this case, the lattice network may not be sufficient to produce a visible agglutination; the high concentration of antibody binds to antigenic sites in such a manner that cross-linking of the complexes cannot occur (Figure 21). This may result in a false-negative reaction (no agglutination) when higher quantities of HIV antibody in the serum are present.

In the case of antibody excess and the occurrence of the prozone phenomenon, antibody in the sample can be detected if the sample is diluted and retested. Dilution of the serum decreases the concentration of antibody to a point where the optimal concentration of antigen and antibody occurs, and agglutination proceeds. It is important that laboratorians be aware of this phenomenon. Sometimes an equivocal reaction (+/–) may occur with the undiluted sample, suggesting that prozone may be occurring. Although prozone

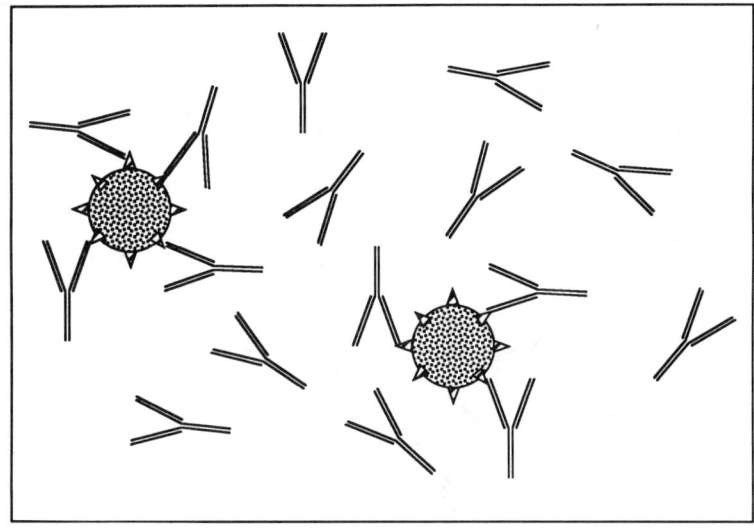

FIGURE 21. Absence of agglutination due to prozone (antibody excess).

reactions do occur, the initial dilutions of the test sample recommended by the manufacturer usually prevent the prozone phenomenon from presenting a problem.

Agglutination assays are very easy to perform and require no wash procedures. The sole requirement is an initial dilution of serum, addition of coated particles, mixing, and incubation, usually at room temperature. Reactions are read macroscopically, or with the aid of a magnifying mirror. A control (noncoated or unsensitized particles) should be used to detect agglutination due to reactions against the particles themselves. If agglutination is detected in both the control and test, the serum must be retested following adsorption of the serum with uncoated particles to remove these nonspecific agglutinins. Weak agglutinating patterns are sometimes difficult to observe; therefore, all tests must be read very carefully, usually under a high intensity lamp. Typical patterns produced in an agglutination assay are shown in Figure 22.

An HIV antibody technique (SimpliRed) was recently introduced that is performed on whole blood and utilizes the principle of autologous RBC agglutination. In this method, a single reagent is added to whole blood from the individual being tested and the presence of antibodies to HIV are demonstrated by the agglutination of the red cells. The reagent consists of an antibody directed toward common RBC antigens, and is complexed with a synthetic peptide gp41 viral antigen. Therefore, the subagglutinating quantity of antibody attaches to the red cells, and will bring about agglutination of the cells if HIV anti-gp41 antibodies combine with the viral antigen (Figure 23).

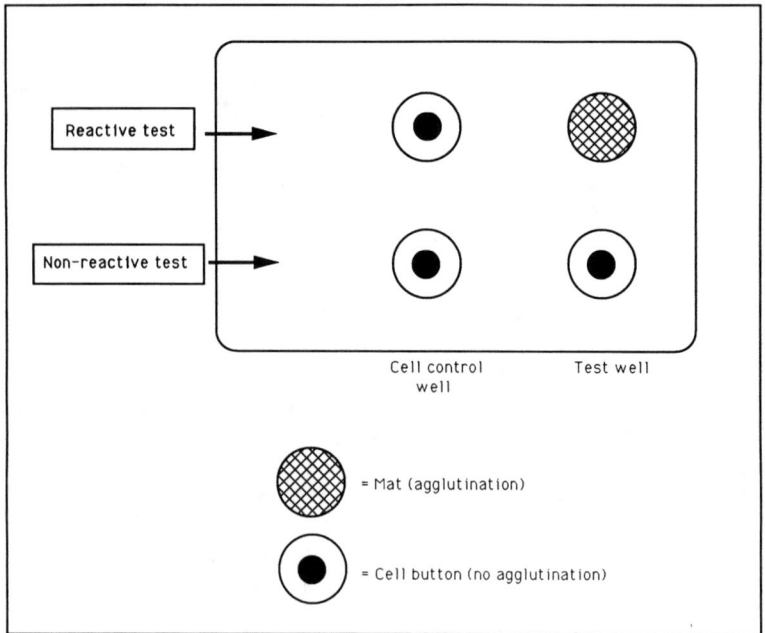

FIGURE 22. Schematic illustrating results of a passive hemagglutination assay test.

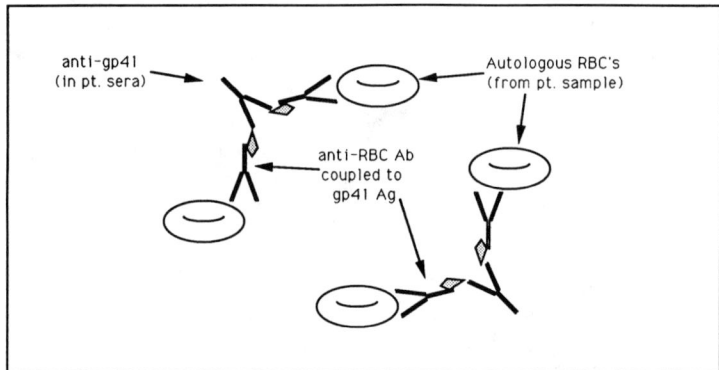

FIGURE 23. Principle of the autologous RBC agglutination test (SimpliRed).

C. Dot-Blot Assays

Dot-blot assays are rapid and easy to perform, but they are also expensive. Many yield results within 5 or 10 min, and some in as little as 3 min.

HIV assays that incorporate paper or nitrocellulose as the solid support are usually referred to as dot-blot assays. In these assays the antigen is passively blotted (absorbed) onto the support. Most often, the antigen is blotted as a small circle (dot), and is usually a recombinant or synthetic peptide antigen. A plastic device often holds the solid support and contains absorbent pads under the paper to collect the serum and reagents. Recently, a similar assay has become available that incorporates antigen-coated microparticles that become trapped within a membrane. These microparticles are microscopic and are very efficient in carrying large quantities of the antigen due to their large surface area. In this assay (Genie and MUREX tests), the reaction occurs on the microparticles that are subsequently collected on the paper so that the resultant color reaction can be visualized.

In these tests, anti-human immunoglobulin conjugates attached to enzymes are commonly used to bind to the patient's antibody. The addition of a suitable substrate produces a color on the paper. Some manufacturers produce tests (e.g., the HIVCHEK and PATH dipstick methods) that use a colloidal gold dye instead of a substrate, and this is conjugated to a substance called protein A. Protein A is derived from the cell walls of the bacterium *Staphylococcus*, and has the ability to bind to most human IgG; therefore, it will attach to the patient's antibody. Because the dye is conjugated to this substance, a colored dot will appear on the paper if antibody is present in the sample.

Some of these dot-blot assays contain a control dot indicating that the test is working properly, that all reagents are good, and that all have been added in the proper order. This control is an anti-human immunoglobulin on the solid support that should bind human immunoglobulin and subsequently the conjugate (or protein A) to produce a color reaction. If this control does not result in the expected result, the test is not performing properly. Recently, dot-blot assays have become available that have separate dots for HIV-1 and HIV-2 (see Chapter 5, Section III. C, HIV-1/HIV-2 "Combination Assays"). Figure 37 (Chapter 5) illustrates a typical reaction that occurs in the dot-blot assays.

Dot-blot assays are convenient to use and can be a valuable asset in certain testing situations such as emergency rooms, blood banks, or autopsy rooms; however, if large numbers of sera are to be tested, their advantage as rapid assays is lost. Some of these tests can be stored at room temperature, and none require instrumentation. Rapid assays are good tests, but because of their expense, their use may be limited. The FDA currently licenses only one rapid assay, the Recombigen latex agglutination test.

D. Other Types of Screening Assays

Recently, Abbott Laboratories introduced an assay that detects HIV antibodies based on a fluorometric microparticle principle. In this assay, which utilizes an instrument known as the IMx, microparticles coated with specific HIV antigens are incubated with serum, and subsequently reacted with a conjugate labeled with a fluorochrome. The microparticles are then trapped in a membrane and the fluorescence that is emitted by the conjugate is detected by a fluorometer. The method is semiautomated and the reaction is based on "in-solution kinetics". Since the microparticles are kept in solution during

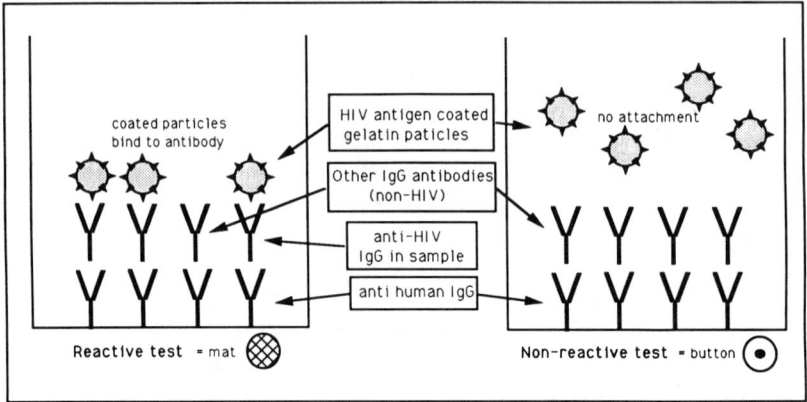

FIGURE 24. The principle of the IgG antibody capture particle-adherence test (GACPAT).

combination of antigen and antibody, the reaction is much faster; the total time is about 15 min. The principle is very similar to the ELISAs, but here the fluorochrome replaces the enzyme and substrate.

Another HIV assay has been developed that combines the principles of two different tests the GACPAT (Wellcome) (IgG antibody capture particle-adherence test). In this test, a microtiter plate is first coated with a rabbit antibody to human IgG that will capture human immunoglobulin of the IgG isotype. When the test serum is added, all IgG antibodies will bind (including anti-HIV IgG). Following incubation and washing, a suspension of HIV-coated gelatin particles is added and allowed to incubate overnight. When anti-HIV is present in the sample the particles will adhere to the well surface rather than settle into a tight button in the bottom of the well. This assay has been reported to yield excellent results when testing unconcentrated urine specimens. Figure 24 depicts the principle of the GACPAT.

E. Recombinant and Synthetic Peptide Antigen-Based Tests

First generation assays based on viral lysate derived antigens are good tests but generally lack specificity (i.e., produce some false-positive reactions). This is due to the presence of contaminating components derived from the cells in which the virus was propagated. If an unknown sample contains antibodies that react with these contaminating components (e.g., antibodies against HLA antigens), a false-positive reaction may occur. In addition, viral lysate tests are usually not 100% sensitive, probably because the high sample dilution needed to attain acceptable specificity eliminates the detection of small amounts of antibody. Also, cellular contaminating substances can occupy sites on the solid phase, thereby decreasing the space that is available for optimal quantities of the HIV antigen to attach. With less antigen attached to the solid support, less antibody can be detected and the indicator signal (color) will be low.

Second and third generation assays utilize important antigenic sites, proteins of precisely targeted portions of HIV, and immunodominant epitopes that have a better chance of detecting significant antibodies to HIV. In addition, the use of these antigens has the advantage of being safer, easier, and more reproducible.

Recombinant antigens are produced when a portion of the HIV genome is inserted into a biological vehicle (e.g., bacterial plasmids) resulting in the production of the gene product (antigen) by the modified vehicle. Since these vehicles can be grown easily in culture, they produce large quantities of the antigen. The advantage of this technique is that the viral antigen of interest is the only viral antigen that is produced, since it was encoded by a particular gene. For example, if the gene coding for gp120 were inserted into the vehicle, the antigen gp120 would be produced without any p24, p55, etc. This results in the production of specific antigens that can be used to develop serologic tests with a high specificity (i.e., few false-positives will be produced). In most cases, antigens used in the recombinant antigen-based tests consist of mixtures of two or more of these recombinant antigens, or portions of antigens, in order to detect antibodies to specific components. In addition, these recombinant antigens can be produced in large quantities and without the hazards associated with growing HIV. Although the antigen is relatively pure, there are some contaminants from the vehicle's components (i.e., the vehicle produces its own proteins). The presence of these contaminating proteins may lead to some false-positive reactions; however, efforts are made to partially purify the recombinant viral antigens. In most instances these recombinant antigen-based tests are excellent tests with high specificity.

Synthetic peptide antigens are the most recent attempt to improve HIV testing technology. Tests based on this technology are gaining popularity. Once the amino acid sequence of an antigen has been determined, the antigen can be made in the laboratory by constructing it from a pool of free amino acids using an instrument known as a protein (peptide) synthesizer. These amino acids are combined in the same order as that of the original antigen. Furthermore, portions of antigens (epitopes) can be synthesized to yield antigens that are only reactive with specific antibodies. Synthetic peptides often consist of a chain of only 10 to 40 amino acids. Serologic assays can therefore be developed using these highly specific peptides or mixtures of two or three of these peptides. The assays will only detect antibodies to the antigens (or epitopes) of interest.

Synthetic peptide antigens can be made in large quantities, similar to recombinant antigens. They have the added advantage of being highly reproducible in a lot-to-lot manner. Since they are in a relatively pure form, there is a high signal-to-noise ratio (low background) when tested, and they are essentially devoid of contaminating components (no host cell contaminants). The use of these antigens helps to eliminate the problems of indeterminate and atypical results that may occur due to antibodies to cellular components in viral lysate-based assays. In addition, the problem of insufficient quantities of certain viral antigens when using virion preparations, which can lead to an underestimation of certain antibodies, can be circumvented by the mechanical application of antigen. In the line immunoassays (LIA) (Chapters 4 and 5), controls can be included that will indicate that the tests are being performed correctly, and can be used to help in the grading of reactions. These qualities mean that synthetic peptide-based tests are extremely sensitive and specific, and probably the best choice for a high quality assay. Their specificity is usually adequate to differentiate HIV-1 and HIV-2 antibodies. Because of their exquisite specificity, many individuals consider these tests to be most useful as confirmatory rather than as screening assays. Some disadvantages exist with these assays. First, the high specificity may not allow for detection of common epitopes shared by similar strains of viruses, and hence the assays may not be as useful for detecting all strains of HIV-1 and HIV-2; i.e., they may lack some sensitivity for detecting all infections. However, manufacturers have attempted to maximize the sensitivity by choosing genetically stable immunodominate regions of the antigens and/or combinations of epitopes. Second, synthetic peptides cannot be glycosated, and therefore the antigens may not parallel the native antigens. However, the recent introduction of conformationally dependent epitopes may address this concern.

III. URINE AND SALIVA TESTS

Within the last 2 years, assays have been developed that use body fluids other than serum for detecting antibodies to HIV. Saliva and urine have been chosen since they are easily collected and do not require an invasive technique for collection. These assays could be valuable for use in certain countries in which needles are still reused for blood drawing. This would help to reduce the risk for infectious diseases such as hepatitis and AIDS. In addition, the collection of saliva and urine is much less expensive than the collection of blood, which requires specific blood drawing equipment. Although urine and saliva HIV tests offer many advantages, some problems have been associated with the development of these assays. Because serological assays have previously lacked sensitivity adequate to consistently detect antibodies in certain body fluids, one finds reluctance among researchers concerning their use. With the development of newer assays such as ELISAs, sensitivities have dramatically increased and detection of antibodies in fluids such as urine, saliva, and breast milk has become possible.

It is now clear that antibodies to HIV are present in saliva from infected individuals, and can be detected with routine screening and confirmatory tests. The isotype of antibody is secretory IgA, and therefore, conjugates must incorporate an anti-human IgA. The synthesis of antibodies in the gastrointestinal (GI) tract is often independent of the systemic immune response and therefore one may find different titers of antibodies and different isotypes of antibody in saliva vs. serum. IgG class antibodies to viral antigens can also be detected in whole saliva; they are derived both from local synthesis and from transudation of serum from ulcers and from gum tissue around teeth. Antibodies to both core and envelope antigens of HIV have been detected in whole and parotid saliva from infected individuals. Recently, there has been much interest in the detection of salivary anti-HIV IgA in newborns. It is believed, but not confirmed, that the presence of secretory IgA in the saliva of newborns may indicate that the infant is truly infected with HIV. This is important, since it is currently difficult to determine if serum antibody in the newborn is derived from the mother (maternal antibody). However, the detection of antibody in saliva is presently not performed routinely, and further investigation into the utility of detecting HIV antibody in saliva is necessary.

The development of tests to detect antibodies in urine has not been attempted in the past, probably because antibodies were considered too large (molecular weight) to pass through the kidney glomerulus, unless significant kidney damage had occurred. In addition, the pH of the urine was thought to interfere with pH-dependent immunological assays. Recently, serological assays have detected HIV antibodies in urine. However, the sensitivities of the tests have thus far proved to be less than optimal. In general, the tests can only detect antibodies in about 92 to 95% of infected individuals, with similar levels of specificity. Both ELISA and Western blot (WB) assays have been used on unconcentrated urine. Many commercial companies are actively pursuing the development of these assays, and it may not be long before they are used routinely in laboratories. Some recent publications have indicated that the performance of tests using urine is approaching that of tests using serum samples. They have also indicated that the antibodies detected in urine are intact IgG molecules, not just antibody fragments.

In conclusion, HIV antibody tests using urine and saliva are producing encouraging results. As technology proceeds, these fluids may become the choice for the detection of HIV infection.

IV. INTERPRETATION OF SCREENING TEST RESULTS

Results generated by screening tests are generally easy to interpret. Each manufacturer of test kits has devised a method of calculation that produces a cutoff value around which test samples can be classified as reactive or nonreactive. This calculated cutoff can be determined based on an average of the negative controls multiplied by a factor, or based on the relationship of the means of the positive and negative controls. Regardless of the exact method to determine the cutoff, the manufacturer has optimized the determination of the cutoff value in order for the test to be capable of identifying most of the sera from infected individuals and to classify as nonreactive samples that have a low degree of reactivity and are from noninfected individuals. In other words, sensitivity and specificity are maximized by the calculated cutoff for each kit.

In most instances, a test result is simply compared to the calculated cutoff and a decision made as to whether it is reactive or nonreactive. For the indirect assays, values above the cutoff are reactive, while for the competitive assays values below the cutoff are reactive. In agglutination and dot-blot assays, results are usually read visually and an interpretation must be made subjectively. Recently, automated readers for the latter tests have been introduced.

Numerous studies have shown that screening tests are very sensitive for detecting antibodies in infected individuals; however, many studies have also determined that they do not possess 100% sensitivity. Individuals who are at a stage of early infection may not have levels of antibodies that will produce a reactive value. It must be understood that all infected individuals will begin with low levels of antibody and gradually the titer will increase as the infection progresses. Therefore, some laboratories will institute a means of identifying persons who may be at this early stage of infection. This is known as "gray zone" identification (Chapter 8). Whether the gray zone method is used, it is important to be vigilant and recognize negative values that are close to the cutoff. In the majority of these cases, the near-cutoff values will be reproducible, and a final interpretation is difficult. It is reasonable to expect that samples that react near the cutoff may also repeat as reactive. Some variability exists within each test (intratest variability), between different lots of tests (interlot variations), and between tests from different manufacturers (interkit variations). If a sample that reacts near the cutoff were repeated 100 times, it is possible that it will be reactive 50 times and nonreactive the remaining 50 times. It is not reasonable or recommended that a sample be tested more than the suggested three times, however, when borderline reactive samples are identified they should be tested by other methods in an effort to resolve the true status. In addition, a subsequent sample should be collected from the individual several weeks later to determine if seroconversion has occurred (see Chapter 8, Section IV.E).

It is well known that errors in the testing process occur. As a means to minimize technical errors, any reactive result is repeated in duplicate. This will generally identify such technical errors as carryover from adjacent samples, the pipetting of wrong samples, etc. It is the practice in some laboratories that a different technologist perform the repeat testing in a blinded fashion to ensure that the result is truly reactive and not a technologist-dependent error. Most initially reactive samples do repeat, although a significant number do not. Therefore, a result must not be interpreted and reported until repeat testing is performed and the screening result verified.

Even following repeat testing, reactive results may occur with individuals who are not infected. Similar to the false-negative results produced when an individual is in an early stage of infection, these false-positive results may not be due to technical error, but rather to a reason related to the sample. These would therefore be considered as biologic false-positives. Reasons for biologic false-positives are not known in the majority of cases; however, if an individual possesses antibodies to HLA antigens or other components of the cells in which the virus was grown, a reactive result may be produced. In addition some persons with autoimmune diseases such as lupus and rheumatoid arthritis may produce reactive results without being infected with HIV. Therefore, reactive results by screening tests must be interpreted with caution, and a supplemental test performed to verify the true status of the individual.

Although it is not always true, it is generally noted that samples that produce very high O.D./cutoff (C.O.) values (see Chapter 8) are more likely to represent true infection; i.e., most samples with high O.D. values will ultimately be confirmed by another test. Therefore, the O.D./C.O. value can give an indication as to whether the sample will most likely be confirmed. This should not be used as a means to interpret a result for final analysis, since it has been noted by the American Red Cross that a small number of samples with O.D./C.O. ratios of 1.0 to 1.2 by ELISA have been confirmed as positive by WB. Most importantly, samples that produce O.D. values near the cutoff should be tested for the presence of antibodies to HIV-2. Some sera from persons infected with HIV-2 will not produce strongly reactive results using HIV-1 screening tests (Chapter 5).

It must be emphasized that screening tests are not perfect and the results of these tests may not be 100% accurate. Inaccuracies occur, whether due to technical, clerical, or some inherent fault of the assay. Therefore, all results should be interpreted with caution, reported only after the results have been thoroughly reviewed and after all testing has been subjected to a rigorous quality assurance (QA) program (Chapter 8).

V. SELECTION OF A SCREENING TEST

One of the most commonly asked questions is "which screening test is the best?". At present no one single test stands above all others. In addition no combination of tests can be the most appropriate for all testing situations. Tests offer different advantages and characteristics that must be considered in any particular testing situation. Some screening tests are very specific, while others offer greater sensitivity; some are rapid but expensive; some require less expertise and equipment, but are cumbersome to use if large numbers of sera are to be tested. Each laboratory must decide which test to adopt, depending on their needs and legal or certification requirements. As stated previously, no test is perfect, but most offer equivalent sensitivities and specificities.

The choice of a test should not be determined on the basis of the manufacturer's claims, or even based on the studies of others. Claims and studies should be used as a guide in selection, but adoption should be made on the basis of test performance in one's own laboratory setting and using the population of samples where the test will ultimately be used (see Chapter 10). The performance of a particular test is known to vary between laboratories, probably due to differences in such variables as the quality of distilled water, ambient temperatures, atmospheric pressures, instrumentation, and laboratory technique.

A wide variety of factors are to be considered in test selection. Some testing situations may dictate that a test be used that requires no instrumentation, while other situations may demand a fully automated test. The shelf life of reagents and the storage conditions must also be considered, depending on availability and the number of tests to be performed. In certain situations (such as blood donor screening) a test possessing the highest sensitivity (to detect all positives) must be selected to ensure that the blood supply is safe. In other situations such as private physician offices where therapy may be instituted for a patient, a test must have a high positive predictive value (Chapter 7). Some laboratories may be able to afford the presumably better, more expensive tests (such as recombinant and synthetic peptide), while others cannot. If HIV-2 infection is suspected in a geographic region, the use of a combination test should be considered. The characteristics of the population to be tested also must be considered when adopting a test.

The choice of a test depends on the situation and the need. With such a wide variety of tests available, each laboratory should be able to find an appropriate test to produce reliable results in their testing situation. It is fortunate that so many commercial companies produce such a variety of HIV tests using different technologies. As these companies compete, technology is driven forward, and better tests become available. Perhaps we will eventually have tests that are essentially perfect for every testing situation.

VI. COMMERCIALLY AVAILABLE SCREENING TESTS

Because such a large number of HIV screening tests are available, it is not possible to list all of them. Some tests are not available for use in all countries. For example, a screening test that is not licensed by the FDA cannot be used in the U.S. Some of these tests are available on a research use-only basis, while others are not even available in some countries. Research use-only tests cannot be used for screening blood and their results cannot be used for the management of a patient.

Table 1 offers a partial list of some of the tests that are available to screen for antibodies to HIV-1 as of early 1992. This list provides some basic characteristics of each test, and indicates which tests are currently licensed by the FDA for use in the U.S.. Organizations such as WHO may be able to provide more extensive information on the characteristics of the available HIV screening assays.

TABLE 1
A Partial List of Available Screening Tests to Detect Antibodies to HIV-1[a]

Name of test	Manufacturer (alphabetical)	Type of Test	Antigen
HIV ELISAs			
For the Detection of Antibody to HIV-1			
Abbott HIV AB[b]	Abbott Laboratories	I	L
Abbott Recombinant HIV EIA	Abbott Laboratories	I	RP
Enzygnost Anti-HIV micro	Behringwerke	C	L
Recombigen HIV-1 EIA[b]	Cambridge Biotech	I	RP
Karpas Cell Test HIV I & II	Cambridge Virucells	I	IC
Retro-Tek HIV ELISA[b]	Cellular Products	I	L
DuPont HIV-1 Recombinant	DuPont de Nemours	I	RP
DuPont ELISA[b]	DuPont de Nemours	I	L
ELAVIA I	Diagnostics Pasteur/Sanofi	I	L
HIV-1 ELISA	Diagnostic Biotechnology	I	L
LAV EIA[b]	Genetic Systems/Sanofi	I	L
REC VIH-KCO1	Heber Biotec	I	RP
Select HIV	IAF Biochem/Coulter	I	SP
HIV-1 env Peptide EIA	Labsystems	I	SP
Ortho HIV ELISA System[b]	Ortho Diagnostic Systems	I	L
Vironostika Anti HIV-1[b]	Organon Teknika	I	L
Vironostika Uni-Form HIV-1	Organon Teknika	C	L
HIV-TEK G	Sorin Biomedice	I	L
MicroTrak HIV-1 EIA	Syva	I	RP
UBI ELISA[b]	United Biomedical International	I	SP
Wellcozyme Recombinant	Wellcome Diagnostics	C	RP
For the Detection of Antibody to HIV-1 and HIV-2			
IMx	Abbott Laboratories	F	RP
Rec. HIV-1/HIV-2 3rd generation[b]	Abbott Laboratories	S	RP
Enzygnost Anti HIV-1 & 2	Behringwerke	I	SP
Biotest Anti-HIV-1/2	Biotest Diagnostics	I	RP
HIV-1/HIV-2 Modul-Test	Biochrom	I	SP
Peptide HIV ELISA	Cal-Tech Diagnostics	I	SP
HIV (1 + 2)	Clonatec	I	SP
ELAVIA mixT	Diagnostics Pasteur/Sanofi	I	L
Rapid ELAVIA mixT	Diagnostics Pasteur/Sanofi	I	L
Genelavia mixT	Diagnostics Pasteur/Sanofi	I	SP/RP
DuPont HIV-1/HIV-2 ELISA	DuPont de Nemours	I	SP/RP
HIV-1/HIV-2 EIA[b]	Genetic Systems/Sanofi	I	L
Anti-HIV-1/HIV-2 EIA	Hoffmann-LaRoche	I	RP
Human HIV 1 & 2	Human	I	RP
Detect-HIV-1/2	IAF Biochem/Coulter	I	SP

TABLE 1 (Continued)
A Partial List of Available Screening Tests to Detect Antibodies to HIV-1[a]

Name of test	Manufacturer (alphabetical)	Type of Test	Antigen

For the Detection of Antibody to HIV-1 and HIV-2

Name of test	Manufacturer	Test	Antigen
Select HIV-1/2 Diff	IAF Biochem/Coulter	I	SP
Innotest HIV-1/2	Innogenetics	I	SP/RP
Vironostika HIV-MixT	Organon Teknika	I	L/SP
Wellcozyme HIV-1 + 2	Wellcome Diagnostics	S	SP/RP
GACELISA	Wellcome Diagnostics	S	RP

For the Detection of Antibodies to HIV-1, HIV-2, HTLV-I, and HTLV-II

Name of test	Manufacturer	Test	Antigen
Bioelisa HIV-1 + 2, HTLV-1 + 2	Biokit Ltd.	I	SP
Detect Plus	IAF Biochem International	I	SP

HIV Rapid and Simple Assays

For the Detection of Antibody to HIV-1

Name of test	Manufacturer	Test	Antigen
Retro Cell	Abbott Laboratories	A	L
SimpliRed	Agen	A	SP
Recombigen HIV-LA[b]	Cambridge Biotech	A	RP
HIVCHEK	DuPont de Nemours	Dot	RP
Serodia HIV	Fujirebio	A	L
SUDS	MUREX	Dot	L/SP
PATH HIV dipstick	PATH	Dot	SP
Immunocomb	PBS Organics	Dot	SP
Serion Immuno Tab HIV-1	Serion Immunodiagnostics	Dot	L
GACPAT	PHLS Laboratories	A	L

For the Detection of Antibody to HIV-1 and HIV-2

Name of test	Manufacturer	Test	Antigen
Test Pack HIV-1/HIV-2 AB	Abbott Laboratories	Dot	RP
Recombinant Rap Test Dev.	Cambridge Biotech	DOT	RP
Rapid HIV-1/2	Clonatec	Dot	SP
HIV-SPOT	Diagnostic Biotechnology	Dot	SP/RP
HIVCHEK 1 + 2	DuPont de Nemours	Dot	SP/RP
Genie HIV-1/2	Genetic Systems/Sanofi	Dot	SP
SUDS 1 + 2	MUREX	Dot	L/SP
Immunocomb Bi-Spot	PBS Organics	Dot	SP
Recodot	Waldheim Pharmazeutika	DOT	RP

Key: C = competitive ELISA, L = lysate, RP = recombinant protein, LIA = line immunoassay, CAP = capture assay, A = agglutination assay, Dot = dot-blot or microparticle assay, SP = synthetic peptide, I = indirect ELISA, S = sandwich ELISA, IC = infected cells, F = fluorescence.

[a] May not include all available tests.
[b] Licensed by the FDA.

REFERENCES

AIDS 89 Summary. A Practical Synopsis of the Fifth International Conference, Philadelphia Sciences Group, Philadelphia, 1989, 1.

Archibald, D. W., Johnson, J. P., Nair, P., Alger, L. S., Herbert, C. A., Davis, E., and Hines, S. E., Detection of salivary immunoglobulin A antibodies to HIV-1 in infants and children, *AIDS,* 4, 417, 1990.

Burke, D. S., Laboratory diagnosis of human immunodeficiency virus infection, *Clin. Lab. Med.,* 9, 369, 1989.

Committee on Human Retrovirus Testing, Fifth Consensus Conference on Testing for Human Retroviruses: Report and Recommendations, Kansas City, MO, 1990, 5.

Connell, J. A., Parry, J. V., Mortimer, P. P., Duncan, R. J. S., McLean, K. A., Johnson, A. M., Hambling, M. H., Barbara, J., and Farrington, C. P., Preliminary report: accurate assays for anti-HIV in urine, *Lancet,* 335, 1366, 1990.

Constantine, N. T., Farag, M. M., and Watts, D. M., Evaluation of a synthetic peptide-based assay and a rapid dot-blot recombinant test for the detection of antibodies to HIV-1 and HIV-2, *J. Egypt Publ. Health Assoc.,* 115, 1, 1990.

Constantine, N. T., Fox, E., Abatte, E. A., and Woody, J. N., Diagnostic usefulness of five screening assays for human immunodeficiency virus in an East African city where prevalence of infection is low, *AIDS,* 31, 313, 1989.

Desai, S., Bates, H., and Michalski, F. J., Detection of antibody to HIV-1 in urine, *Lancet,* 337, 183, 1991.

Epstein, J., Gregg, R., Saah, A., Phair, J., Fahey, J., Rinaldo, C., and Polk, B., Western blot analysis of serial IgG, IgM, IgA responses to the human immunodeficiency virus (HIV) in recent seroconverters, presented at the 3rd Int. Conf. AIDS, Washington, D.C., 1987, 131.

Epstein, J. S., Sensitivity and consistency of screening tests for antibodies to human immunodeficiency virus type 1, *Transfusion,* 31, 388, 1991.

Report of the WHO Meeting on Criteria for the Evaluation and Standardization of Diagnostic Tests for the Detection of HIV Antibody, Stockholm, GPA/BMR/88.1, World Health Organization, Geneva, 1987.

Operational Characteristics of Commercially Available Assays to Determine Antibodies to HIV-1, GPA/BMR/90.1, World Health Organization, Geneva, March 1989.

Operational Characteristics of Commercially Available Assays to Determine Antibodies to HIV-1 and/or HIV-2, Report 2, GPA/BMR/90.1, World Health Organization, Geneva, April 1990.

Operational Characteristics of Commercially Available Assays to Determine Antibodies to HIV-1 and/or HIV-2, Report 4, GPA/BMR/91.6, World Health Organization, Geneva, October 1991.

Gnann, J. W., McCormick, J. B., and Mitchell, S. W., Synthetic peptide immunoassay distinguishes HIV type 1 and HIV type 2 infections, *Science,* 237, 1346, 1987.

Griner, P. F., Mayewski, R. J., Mushin, A. I., and Greenland, P., Selection and interpretation of diagnostic tests and procedures, *Ann. Int. Med.,* 94, 557, 1981.

Kelly, E. B., The search for a rapid HIV tests, *AIDS Patient Care,* February 1990, 25.

Miller, L. E., Ludke, H. R., Peacock, J. E., and Tomar, R. H., *Manual of Laboratory Immunology,* 2nd ed., Lea & Febiger, Philadelphia, 1991.

Mitchell, S. and Mboup, S., HIV testing, in *The Handbook for AIDS Prevention in Africa,* Family Health International, 1990, 20.

Sloand, E. M., Pitt, E., Chiarello, R. J., and Nemo, G. J., HIV testing state of the art, *JAMA,* 266, 2861, 1991.

Sun, D., Archiblad, D. W., and Furth, P. A., Variation of secretory antibodies in parotid saliva to human immunodeficiency virus type 1 with HIV-1 disease stage, *AIDS Res. Human Retroviruses,* 6, 933, 1990.

Turgeon, M. L., *Immunology and Serology in Laboratory Medicine,* C.V. Mosby, St. Louis, 1990.

Van de Perre, P., Hitmana, D. G., and Lepage, P., Human immunodeficiency virus antibodies of IgG, IgA, and IgM subclasses in milk of seropositive mothers, *J. Pediatr.,* 113, 1039, 1988.

Voller, A., Bidwell, D. E., and Bartlett, A., *The Enzyme Linked Immunosorbent Assay (ELISA),* Zoological Society of London, London, 1979.

Weiblen, B. J., Schumacher, R. T., Garrett, P. E., and Hoff, R., IgA and IgM human immunodeficiency virus antibodies in weakly reactive or false-negative blood donors, *Transfusion,* 31, 397, 1991.

Zhang, X., Constantine, N. T., Bansal, J., Callahan, J. D., and Marsiglia, V. C., Evaluation of a new generation synthetic peptide combination assay designed to detect antibodies to HIV-1, HIV-2, HTLV-I, and HTLV-II simultaneously, *J. Med. Virol.,* in press, 1992.

Chapter 4

SUPPLEMENTAL TESTS FOR HIV-1 INFECTION

I. INTRODUCTION AND USE

Although HIV screening tests are adequate for identifying most HIV infections, they generally lack a sufficiently high degree of specificity. This means these screening tests will produce some false-positive results (biologic false-positives). It should be remembered that screening tests were designed to protect the blood supply, and their sensitivity is adequate to accomplish that purpose. However, since their specificity may be less than optimal, results must be verified to substantiate HIV infection.

As discussed previously, most widely used screening tests are based on viral lysate technology, and consequently contain host cell contaminants. Because of the occurrence of false-positives, the results of the screening assays cannot, and must not, be considered conclusive. This is particularly evident when testing certain low-risk populations, where the screening tests exhibit a poor positive predictive value (Chapter 7).

Confirmatory (or supplemental) testing is highly recommended for samples that are repeatedly reactive for antibodies to HIV by screening assays. In some countries, confirmatory testing consists of retesting the repeatedly reactive sample using a second screening assay. The principle of the second screening assay should be different than that of the first. For example, if an indirect ELISA were repeatedly reactive initially, then a repeatedly reactive result by a competitive ELISA is sometimes considered as the final result "positive for HIV". The objective in determining the best combination of assays is to detect all false-positive sera as well as all true positives.

Most protocols, however, require the use of very specific assays such as the Western blot (WB), indirect fluorescent antibody assay (IFA), or the radioimmunoprecipitation assay (RIPA) to verify screening test results. These tests are extremely specific, and if performed and interpreted correctly, should not produce biologic false-positive results. They are, however, more laborious and more expensive than screening assays.

As discussed below, confirmatory assays may not always produce definitive results (i.e., do not always yield a positive or negative result); therefore, additional testing may be required before a conclusion can be made. The primary purpose of confirmatory tests is to ensure that individuals who test reactive by screening assays are not incorrectly identified as being HIV infected.

II. PRINCIPLES OF TESTS

A. Western Blot

The WB is probably the most widely accepted confirmatory assay for the detection of antibodies to the retroviruses. The technique was developed many years ago and is recognized as a very specific technique for detecting antibody to many infectious agents. Many authorities consider it the "gold standard" for validation of HIV results. The WB is at least as sensitive as the screening assays, but is expensive, labor-intensive, and therefore is inappropriate as a screening tool.

The WB technique owes its exquisite specificity to two factors: component separation and component concentration. As the mixture of viral components is separated into specific "bands", each component becomes relatively pure. In addition, the separation of antigens allows for the identification of specific antibodies to each of the viral antigens. In principle, it is conducted in three parts:

1. Separation of viral lysate antigens in relation to their molecular weights by sodium dodecyl sulfate-polyacrylamide gel electrophoresis (SDS-PAGE)
2. Transfer (blotting) of the separated antigens, usually onto nitrocellulose paper
3. Testing of the unknown serum sample directly on the blotted membrane using a technique similar to the ELISA (enzyme-substrate reaction)

Most WB used for detecting infection by the retroviruses are supplied by commercial companies in kit form and only the third part (see list above) is performed in the laboratory. The tests are easy to perform, but they require strict adherence to procedure; most importantly, they require the proper interpretation of results.

Laboratories in which the WB was developed in-house may save money, but usually require facilities to propagate the virus, expensive equipment for electrophoresis and blotting, and usually must contend with the hazards associated with handling live virus. In addition, the use of preblotted strips from the manufacturers has greatly reduced the variability associated with WB preparations, and therefore helps to standardize the technique. In order to produce WB, large quantities of virus must be grown in culture, then chemically treated for disruption and inactivation of the virus. Antigens are separated through a gel in an electric field in the presence of SDS (an anionic detergent that denatures the viral components and yields proteins with net negative charges), then is blotted electrophoretically onto nitrocellulose. In the presence of SDS, the negatively charged proteins migrate in polyacrylamide gel according to their molecular weights. Bands of proteins corresponding to certain molecular weights are thus produced. The low molecular weight components migrate the fastest,

Supplemental Tests for HIV-1 Infection

and hence are found furthest from the point of origin (toward the bottom of the strip at the anode). Conversely, the higher molecular weight components migrate slowly, and stay near the origin (toward the top at the cathode). The HIV-1 viral antigens are therefore separated as follows (from top to bottom): gp160, gp120, p66, p55, p51, gp41, p31, p24, p17, and p15. It is important to remember that nonviral proteins derived from the host cells in which the virus was grown are also present on the nitrocellulose. They can occur in many places, but often are near the mid-molecular weight region (40,000 to 60,000).

As previously mentioned, commercial WB kits are supplied with individual strips of nitrocellulose that already contain the blotted, separated antigens. Each strip is used as a solid phase and an ELISA procedure (indirect) is performed directly on the strip (Figure 25).

To perform the WB, test strips are placed in individual troughs of a plastic tray. In the initial step, each strip is wetted and incubated in the presence of a blocking solution (usually albumin or a milk powder solution). The protein in this solution binds to the strip in areas where there are no components (i.e., no antigens or host cell proteins). Thus, the purpose of this solution is to prevent any non-HIV immunoglobulin in the sample from binding to unoccupied areas of the strip, which would produce a nonspecific reaction when subsequent reagents are added.

Following this blocking step, the test sample is added and incubated for a specified time period; care must be taken to avoid splashing or contamination between troughs during the procedure. The plastic tray is placed on a rocker platform and gently rocked for a specified incubation period. Incubation temperature varies with kit manufacturer, as does incubation time. During incubation, specific antibodies to the viral components will bind to their respective antigens. If the sample contains antibodies to p24 and gp41, for example, there will be binding only to those antigens. Subsequent washing will remove unbound serum components.

Conjugate addition, incubation, and washing follow these steps. Most often, the conjugate is an anti-human immunoglobulin coupled to an enzyme (as in the ELISA tests). Other types of conjugates are often used and may consist of two separate components: an anti-human immunoglobulin coupled to the vitamin biotin (conjugate 1), followed by the addition of a protein (derived from streptococcus) called avidin, which is coupled to an enzyme (conjugate 2). In this latter case, the biotinylated anti-human immunoglobulin binds to the anti-HIV antibody in the sample (if present) and the avidin-enzyme reagent binds to this biotin. The advantage of this biotin-avidin system is an increase in sensitivity (detects smaller quantities of antibody) since many molecules of the second conjugate (avidin-enzyme reagent) can bind to the biotin anti-immunoglobulin reagent, resulting in a strong visual signal that can be detected. An example of this WB conjugate reaction is depicted in Figure 26.

Finally, a precipitable substrate is added to the WB system, resulting in the formation of a colored end-product; a precipitate forms on the nitrocellulose

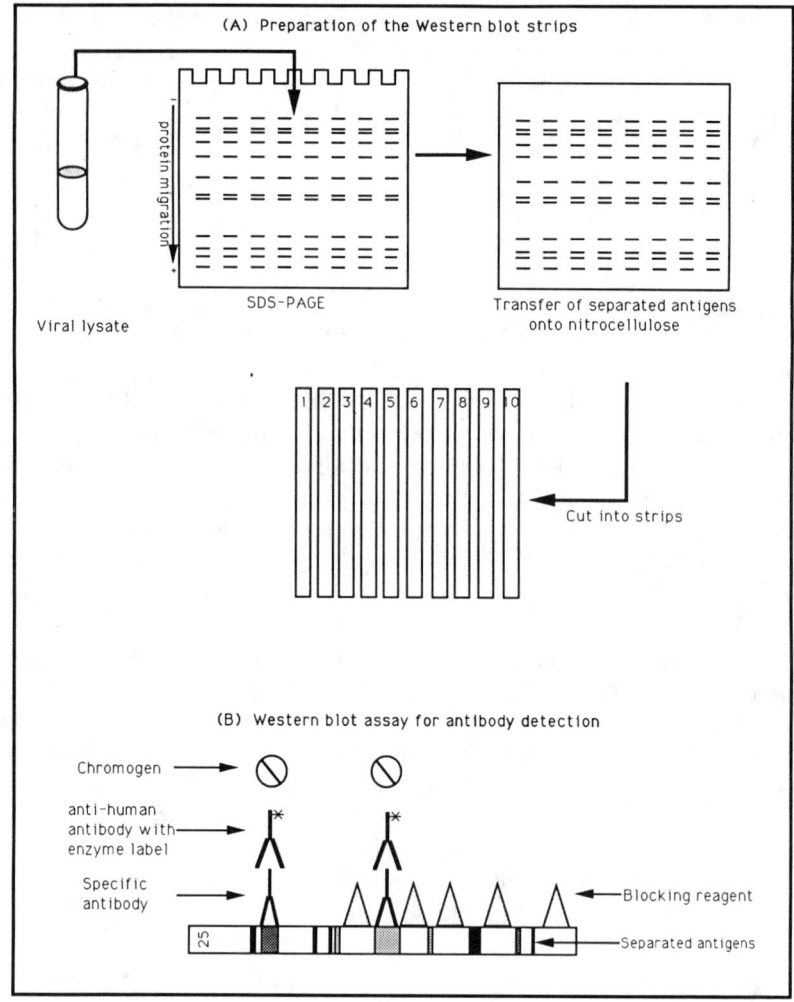

FIGURE 25. Principle of the WB. (A) Performed by manufacturer; (B) performed routinely in laboratory.

strip at the particular sites where specific antibodies have bound. Depending on the particular antibodies in the sample, band profiles will be produced. The type of profile (the combination and intensity of bands that are present) will determine whether the individual is considered to be positive for antibodies to HIV (see Section III). Figure 27 is a schematic representation of positive and negative reactions by a typical WB.

While performing the WB procedure, care must be taken to avoid cross-contamination of test strips during pipetting and washing steps. As with other assays, clean plasticware is essential because chemical contamination can

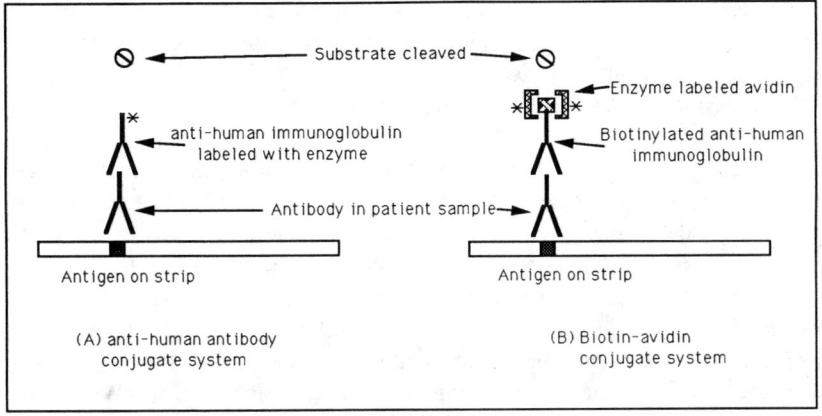

FIGURE 26. The anti-human and biotin-avidin conjugate systems.

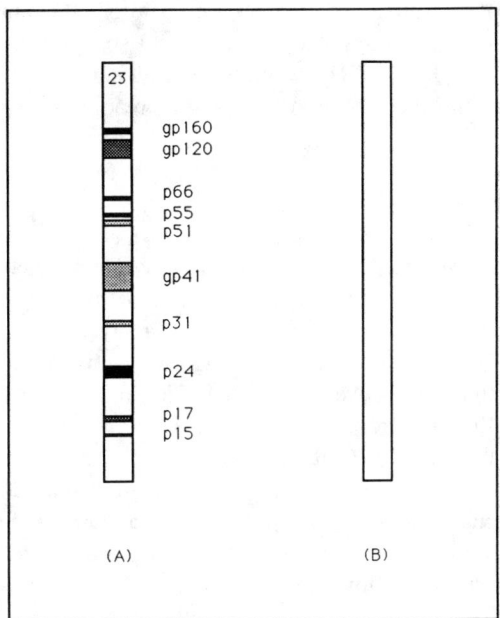

FIGURE 27. Schematic representation of WB results: (A) positive result (reactions to all antigens); (B) negative result.

affect reactions. As with all serological tests, the proper controls must be included with the WB assay. In particular, a weak positive control is used as a baseline for positivity (i.e., it produces the minimum intensity reaction required for a sample to be considered positive). The controls must be run on the same lot of strips as the samples, since the relative position of antigens on strips may vary from lot to lot.

At the completion of each run, reactions should be read immediately (unless otherwise stated by the manufacturer), and then the strip can be dried and stored. If stored, strips should not be taped but enclosed in a plastic binder. Since the bands may fade with time, a written record or photocopy of the reactive bands immediately after completion of the assay must also be included. The interpretation of WB results is discussed in Section III.

Several commercial companies have recently introduced automated WB methods. The System 27 AutoBlot (Diagnostic Biotechnology) is a computer-based, temperature-controlled instrument that has the capability of incubating specimens, adding reagents, and signaling the completion of the WB assay. The unit has a cover, which helps to minimize exposure of the technologist to the samples. Another company has introduced a semiautomated dot-blot assay, very similar to the WB, for determining antibody profiles to HIV. This system, the MATRIX, uses highly purified recombinant-derived DNA proteins on a nitrocellulose-based solid phase to capture specific HIV antibodies in the sample. The test cell employs a series of distinct spots on the solid support, each representing a different HIV antigen. Following a typical enzyme-substrate reaction, the resultant antibody profiles are read by the optical system of the instrument.

B. Line Immunoassays (LIA)

The LIAs are second or third generation assays and have potential for use as supplemental tests. Recombinant proteins and/or synthetic peptides are applied in band patterns on plastic support strips or nitrocellulose strips and tested in a manner similar to the immunoblot assay. Most LIAs utilize the indirect ELISA methodology for visualizing the results (Figure 28).

The major differences between the LIA and the WB are that the antigens are applied, rather than electrophoresed, and are derived from recombinant or synthetic peptide technology rather than viral lysates. In the LIA, optimal quantities of antigens can be applied, resulting in an increased ease of reading and interpretation. In addition, the presence of contaminating substances is minimized, resulting in less distraction and cleaner antigenic profiles. The use of these antigens helps to eliminate the problems of indeterminate and atypical results that may occur due to antibodies to cellular components in viral lysate-based confirmatory assays. In addition the problem of insufficient quantities of certain viral antigens from virion preparations, which can lead to an underestimation of certain antibodies, can be circumvented by the mechanical application of antigen. These assays also allow for the application of antigens from more than one virus, thereby allowing them to act as combination assays and to differentiate infection by HIV-1 and HIV-2 (Chapter 5). The LIAs also contain control lines that indicate that the test is performed properly, and can

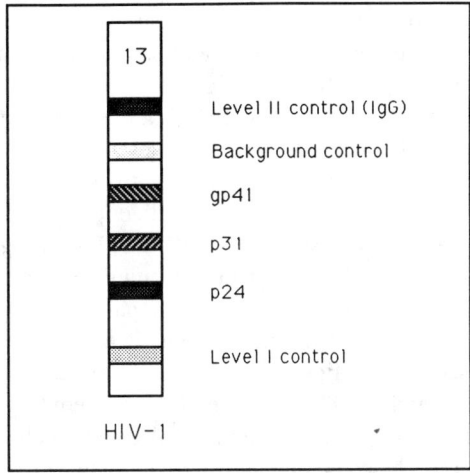

FIGURE 28. An example of an LIA.

act as a guide in estimating the intensity of reactions. In most cases they can be performed in about 2 h (less than most WBs), and can be performed on serum, plasma, and whole blood.

C. Indirect Immunofluorescence Assay (IFA)

The IFA test is used in many laboratories as a confirmatory test for HIV infection, although some laboratories use it as a screening test. The test is usually less expensive than WBs and rapid assays, is simple to perform, and requires much less time to perform than many of the WB assays. The major disadvantages are the requirement of an expensive fluorescence microscope, the absence of a permanent record, and the need for experienced laboratory personnel to read and interpret the results. In addition the test is cumbersome to perform and read if many samples are tested at one time. One HIV IFA test was licensed by the FDA in mid-1992, and can be used to help resolve indeterminate WB results (see Section III.B).

In principle, the IFA is similar to the ELISA, with the major differences being the antigens on the solid support and the indicator system, which is a fluorochrome in the IFA. The solid support consists of a microscope slide containing cells (usually human T lymphocytes) that have been infected with HIV. Also included is a control consisting of noninfected cells. The infected cells and control cells are fixed (attached) onto the slide in circumscribed areas referred to as "wells". Because reactions will take place in these fixed cells, the cells are referred to as substrates (not to be confused with enzyme substrates). The slides containing the substrate are usually packaged with four wells containing infected cells and four wells of uninfected control cells. Alternatively, a single well may contain both infected and noninfected cells. In the former case, nonspecific reactions can be detected in the control wells; while in the latter case, nonspecific reactions would be indicated if all the cells in a well exhibited a reaction (see Section III).

Test serum (usually 10 to 20 µl of sample in diluent) is added to each well using a mechanical pipette. Incubation of the slides, usually for 30 min at 37°C, allows specific antibody (if present) to attach to the viral antigens in the infected cells. The slides are then washed in a bath of phosphate buffered saline (PBS) for 15 min with one or two changes of PBS. Following this, the slides must be dried (by air or with a small fan). Conjugate is added in the same manner as the serum. The conjugate is an anti-human immunoglobulin labeled with a fluorochrome (FITC). The fluorochrome is a substance that will fluoresce when exposed to UV light. This occurs when the molecules of the fluorochrome are excited to a higher energy level and emit light of a different wavelength as they return to the ground state. During incubation (usually 30 min, 37°C), the FITC-labeled conjugate will bind to the anti-HIV (if present in the sample). Another wash and drying step follows to remove any unbound conjugate. Buffered glycerol is added to each well (to decrease the refraction of light) and a coverslip is placed over the entire slide. The test should be read immediately, but can be stored in the dark at 4°C and read the following day; alternatively, the coverslip can be sealed with nail polish and the slides stored refrigerated for days before reading. The slides are examined under UV light with the aid of a fluorescence microscope.

The viral antigens are contained in the cytoplasm of the substrate cells and therefore a positive (reactive) test will exhibit cytoplasmic fluorescence. In contrast to the WB, the IFA will not identify antibodies to specific antigens; a positive reaction only indicates that antibodies to HIV are present. Since lymphocytes have large nuclei that occupy most of the cell, the fluorescence is generally seen as a wedge-shaped area in the cytoplasm, although surface-expressed antigens may produce some membrane fluorescence. The intensity of fluorescence should be graded as 1+ (weak) to 4+ (strong); appropriate serum controls will act as a guide. For a serum to be considered positive for antibodies to HIV, fluorescence should not be present in noninfected control cells.

It is very important to recognize the staining pattern in order to identify fluorescence due possibly to antinuclear, antimitochondrial, or other anticellular components. Autoantibodies from several autoimmune diseases can result in the attachment of these antibodies to the cells, causing difficulty in reading. This can lead to false-positive results (see Section III). A representation of the principle of the IFA method is depicted in Figure 29.

Several procedural notes are worthy of mention. The IFA slides should remain in their sealed packets and should be opened when at room temperature just prior to use; this will prevent condensation from collecting on the cells and causing lysis. Also, care must be taken to prevent any contact of objects (such as pipettes) with the surface of the wells, so that the cells are not disturbed.

Supplemental Tests for HIV-1 Infection

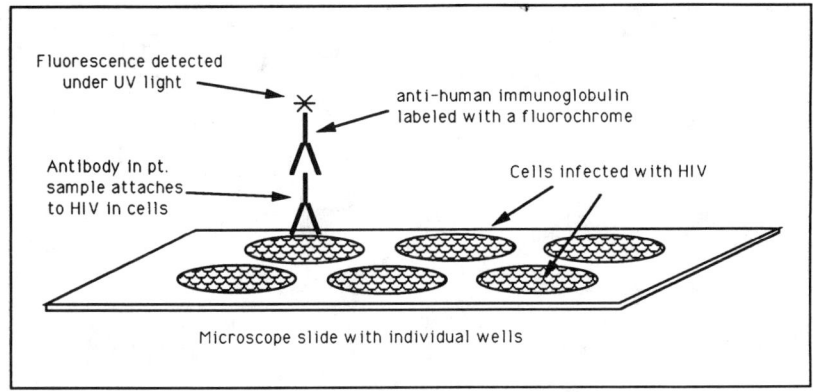

FIGURE 29. The principle of the IFA method.

Most importantly, each slide must be labeled with a number or letter in the event that the slides are inadvertently rearranged during wash steps. Also, the location of each specimen on each slide must be recorded on a worksheet prior to performance of the test. One of the most common problems associated with the performance of the IFA is cross-contamination of samples from one well to another, since the wells are close together. This can be controlled by not adding excessive sample to each well.

D. Radioimmunoprecipitation Assay (RIPA)

The RIPA (or RIP) can be used in conjunction with, or as an alternative to, the WB to confirm HIV infection. Because it is essentially a research technique and requires the use of radioactive substances, it is not used in many clinical laboratories. The RIPA is extremely sensitive, and because the antibody-antigen reaction occurs under nonreducing conditions (no SDS), the test is also very specific.

In the RIPA, certain radiolabeled amino acids (e.g., cysteine) are added to cultures of HIV-infected cells; as the virus replicates it incorporates these labeled protein building blocks into its structural components. Subsequently, the viral particles that have been labeled are collected, disrupted (lysed), and the proteins are solubilized. The resultant viral lysate is added to the serum to be tested. Immune complexes that are produced are subsequently precipitated from solution by protein A, which binds to human immunoglobulins. The labeled precipitate is then electrophoresed by SDS-PAGE and the separated labeled proteins are visualized after exposure to a special film capable of detecting the label (a process called autoradiography).

The labeled antigens in the RIPA are primarily the envelope glycoproteins gp160 and gp120. Both of these bands must be detected in order to classify the serum as antibody positive. The technique is sensitive for detecting antibodies to the envelope proteins, since antibodies to these components are present in

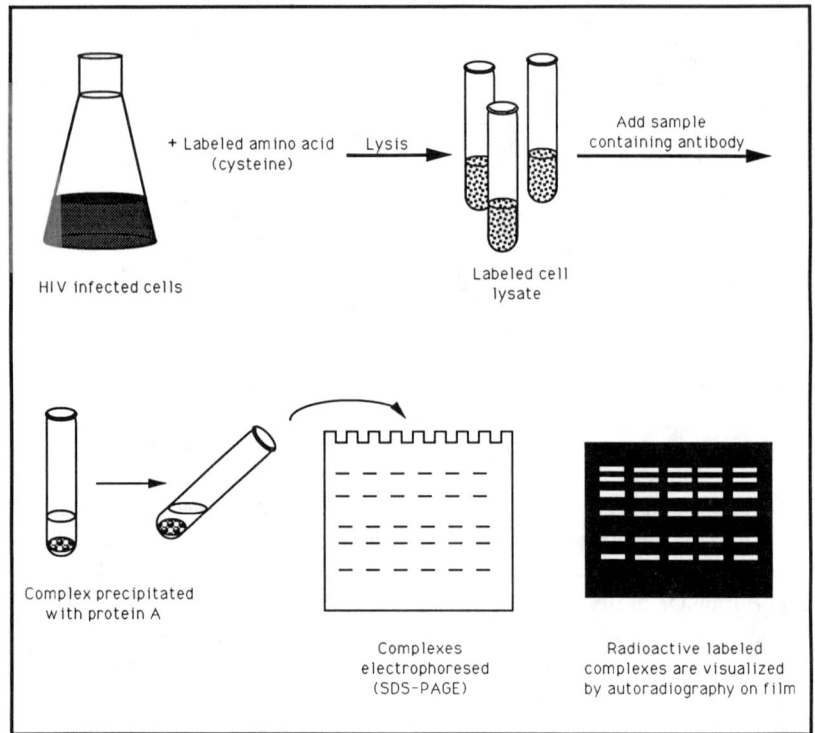

FIGURE 30. The principle of the RIPA.

virtually all HIV infected individuals after seroconversion. Figure 30 indicates the principle of the RIPA.

E. Tests to Detect Circulating Antigen

As noted in Figure 6 (Chapter 2), certain time periods exist during HIV infection that p24 antigen can be demonstrated in serum, thereby verifying infection. However, the antigen test is most useful in detecting antigen in supernatants from cultures that have been inoculated with cells from a patient suspected of being infected (viral culture and isolation). The antigen test is the method of choice for detecting the presence of free antigen in culture, and it is two orders of magnitude more sensitive than the reverse transcriptase assay. The test is most useful for detecting viral antigen in culture fluids because large quantities of antigen are exuded in the supernatant fluid.

The antigen test was not originally designed for the purpose of testing serum, and its use for this purpose is limited. Only about 50 to 60% of AIDS patients, 30 to 40% of ARC patients, and 10% of asymptomatic patients will have p24 antigenemia (antigen in the serum). Furthermore, the degree of antigenemia seems to vary depending on the population tested. For example, several studies have shown that African patients with AIDS have higher levels of antigen than a similar group of individuals in the U.S. Conversely, African patients infected with HIV but asymptomatic, have been shown to have lower levels of antigenemia than a similar group in the U.S. Nevertheless, this test is used occasionally for testing serum, and can be of value for certain situations, including:

- When early infection is suspected and the patient is seronegative
- During late infection, when the patient is usually symptomatic
- To monitor disease progression (a four-fold increase in p24 levels is associated with progression to AIDS within three years) and/or therapy
- To detect infection in a newborn in which diagnosis is difficult due to the presence of maternal antibody (the detection of antigen in a newborn confirms infection and yields information of prognostic significance)

The most probable reason for the lack of sensitivity of the antigen test when testing the serum of AIDS patients is that free antigen (p24) in serum may be complexed with p24 antibody. The test cannot detect complexed antigens. Recently, one manufacturer (Coulter) introduced an immune complex dissociation (ICD) procedure utilizing low pH to dissociate these complexes prior to performing the antigen assay. Using this procedure, an increased sensitivity of the assay was demonstrated, particularly for asymptomatic HIV-infected individuals. This dissociation procedure allows for quantitation of both free p24 antigen and complexed p24 antigen/antibody. This method not only increases the number of antigen-positive individuals who are detected, but also can detect higher amounts of p24 antigen. In addition to the probable presence of complexes, p24 antigen may not be readily detected by this test (particularly in asymptomatic infections) because the viral load (concentration of virus in the individual) may be low. If the virus is not present in high quantity or is not actively replicating, the amount of antigen in serum will be low. The antigen test is relatively insensitive, being able to detect only about 10 pg/ml. This quantity of antigen is considered low, and these levels may not be present in the serum of infected individuals even when the virus is actively replicating. The test can be performed qualitatively or quantitatively using a standard curve generated from standards of known concentration. The quantitative method is useful for monitoring disease progression, with increases in p24 antigen signaling a poor prognosis.

The antigen test can be performed on fluids other than those of culture and serum. Evidence determined from the testing of cerebrospinal fluid (CSF) indicates that many patients with HIV dementia and encephalopathy have detectable antigen in CSF. This is most likely due to active replication of the virus in cerebral tissue.

It is important to realize that this test detects soluble p24 antigen, and does not specifically identify live virus. Simply because a sample is negative for antigen does not indicate that the sample is noninfectious; similarly, the presence of antigen does not by itself confirm that the sample is infectious. The only means available to demonstrate that a sample contains infectious virus is by isolating the virus.

The test to detect HIV antigen employs ELISA technology, but is modified slightly because the presence of antigen, not antibody, is being determined. In the assay, which is an antibody sandwich type, a specific monoclonal antibody to HIV p24 is attached to the solid phase. The test serum is added (diluted in a Triton® X detergent to disrupt virions), and if antigen is present in the serum, the antigen will attach to the monoclonal antibody on the solid phase. Following a wash step, conjugate is added and incubated. The conjugate in this case is an antibody to p24 antigen, coupled to an enzyme; or can be a rabbit anti-HIV followed by a goat anti-rabbit antibody, which is labeled. A typical substrate reaction will subsequently detect the unknown antigen in the sample. Figure 31 depicts the principle of the antigen test.

Although all antigen tests utilize this principle, some require overnight incubation, while others are relatively short (total time of 3 h). However, the tests are subject to false-positive reactions, presumably due to interfering substances and other immune complexes. Therefore, specimens that test reactive in the antigen test must be confirmed, using a more specific method. This is accomplished using an antigen neutralization assay. The principle of neutralization is simple. The sample, presumably containing the antigen, is incubated with a specific anti-p24 antibody (neutralizing reagent). During this incubation, if antigen is present, it will be complexed with the neutralizing antibody. Subsequently, the antigen assay is repeated on this preincubated sample along with an aliquot of the same sample that has not been preincubated with reagent; the O.D. readings are compared. For the sample to be considered confirmed positive for antigen, the O.D. readings of the sample following the neutralization must be reduced by at least 50% compared to the O.D. readings of the non-neutralized aliquot. If this degree of reduction (inhibition) does not occur, the sample is not confirmed for antigen, and the reactivity was probably not due to p24 antigen. Since all of these assays are performed in duplicate or quadruplicate, nearly 1 ml of serum is required to complete the test for a final confirmed result. The test is also relatively expensive. Table 2 lists some of the tests available for detecting p24 antigen.

F. *In Vitro* Antibody Synthesis

Recently, encouraging results have been obtained by inducing HIV-infected cells in culture to produce antibody. This technique is based on the principle that sensitized B lymphocytes will undergo blast transformation and produce specific antibody when exposed to a mitogen such as pokeweed mitogen (PWM), a plant lectin.

In vitro antibody synthesis is accomplished when cells from the patient suspected of being infected are separated, and the mononuclear cells are recovered and placed in cell culture where they are exposed to the mitogen.

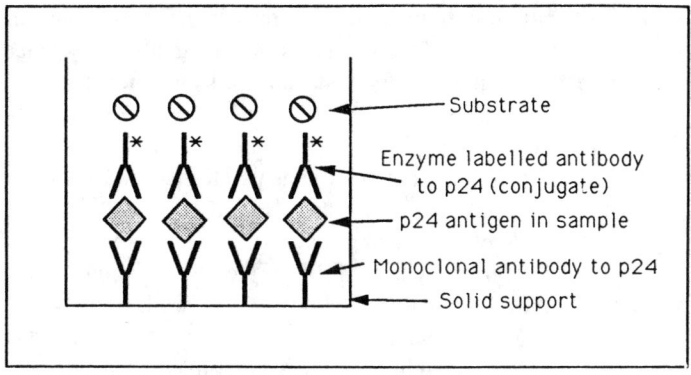

FIGURE 31. Principle of the p24 antigen test.

TABLE 2
Commercially Available p24 Antigen Assays[a]

p24 Antigen EIA[b]	Abbott Laboratories
p24 EIA	Cellular Products
HIV-1 p24 antigen EIA	Coulter
ELAVIA antigen 1	Diagnostics Pasteur/Sanofi
p24 Antigen	DuPont de Nemours
p24 Antigen	Genetic Systems/Sanofi
Vironostika p24 antigen	Organon Teknika
p24 Antigen	Wellcome

[a] May not include all available tests.
[b] Licensed by the FDA.

This mitogen will stimulate all B cells to produce antibody to many different antigens to which the individual had been exposed. If the individual has been infected by HIV, specific HIV antibody will be produced and detected in the culture supernatant by routine HIV antibody assays. This procedure requires about 4 d before antibody can be detected.

In vivo antibody synthesis may be valuable for the detection of antibody during early infection before seroconversion, and in infants born to infected mothers. However, more evaluation is needed before this test is accepted as a standard procedure.

G. Viral Culture

A definitive diagnosis of HIV infection can be made by the use of cell culture to isolate the virus from selected cells, tissues, and body fluids of asymptomatic and symptomatic individuals. A cocultivation system is required in which cultures are prepared from peripheral blood mononuclear cells (PBMC) of HIV seronegative donors. Continuous cell lines derived from neoplastic T cells (HT) also maybe used, but are less susceptible to HIV infection. While

PBMCs isolated from an individual suspected of being infected are the preferred specimen for attempting virus isolation, HIV can also be isolated from CSF, serum or plasma, saliva, cervical secretions, semen, tears, brain tissue, organ biopsies, and breast milk. The isolation success rate for HIV from PBMCs may vary from 100% among symptomatic patients to a substantially lower rate for similar specimens obtained for asymptomatic individuals. Variation in the efficiency of donor seronegative PBMC cultures to consistently isolate HIV is likely to reflect existing levels of virus and/or antibody in the sample, as well as the degree of susceptibility of the PBMC culture to infection. These variables, and the fact that culture procedures are laborious, time consuming, dangerous, and costly have limited the use of this technique for diagnosing HIV infection. In addition, virus isolation also depends on sample size, sample freshness, and the laboratory worker's expertise.

The procedure for the diagnosis of HIV infection by cell culture is to first replicate the virus associated with the test specimen in the susceptible cell suspension or monolayer. A second technique is then employed to demonstrate viral-specific RT activity, or viral p24 antigen in culture medium. The culture assay is performed by the cocultivation of the test specimen, such as the PBMCs, with mitogen-stimulated PBMCs obtained from an HIV seronegative donor. After mitogen stimulation of the PBMCs, the cytokine IL-2 is used to promote growth of these cells. These cocultures are incubated for approximately 1 month. During this period, a sample of culture medium is obtained and replaced with fresh medium once or twice each week. The culture medium is tested for RT activity and/or viral antigen. In addition, freshly mitogen-stimulated donor PBMCs are added once per week to sustain HIV replication. Infection is confirmed by the detection of RT activity or viral antigen in the supernatant, or by demonstrating infection by fluorescence-labeled monoclonal anti-HIV antibodies. A negative culture however, can result from technical problems, decreased ability of the virus to replicate early in the course of infection, or the inability to replicate a virus that may be defective.

Although the detection of viral RT was used initially to confirm HIV infection of cell cultures, this approach has been replaced in most laboratories by the EIA for the detection of viral p24 antigen. The EIA is more sensitive, and the time and cost required to perform this technique are substantially less than that required to perform the RT assay. Also, the latter assay requires radioisotopes, a scintillation counter, and an ultracentrifuge.

While cell culture assays are not used routinely for the diagnosis of HIV infection, the technique offers an alternative approach for clarifying indeterminate serological results. Also, culture assays are useful for isolating HIV for describing the genetic variability among viral isolates. Even though viral isolation is ultimately confirmatory, the method's sensitivity is suboptimal (i.e., some false-negatives occur). Samples must be collected in an anticoagulant such as adedine citrate dextrose (ACD) or heparin, kept at room temperature, and should be tested within 7 h (24 h maximum).

H. Assays to Detect Viral Nucleic Acids

Viral nucleic acid can be detected in HIV-infected cells such as lymphocytes by an *in situ* hybridization technique. This approach takes advantage of the exquisite specificity between complementary strands of nucleic acid. Radio-, enzyme-, or chemiluminescence-labeled HIV RNA is used to probe PBMCs for the presence of proviral DNA. The sensitivity of this technique appears to be very low because HIV infects as few as 1 in 10,000 to 100,000 PBMCs. Therefore, the *in situ* hybridization technique is not considered acceptable for the routine diagnosis of HIV infection.

Recently an assay has been developed (ENZO assay) that incorporates *in situ* hybridization for the colorimetric detection of nucleic acid of HIV. This technique is based on the hybridization procedure performed in 96-well microplates, where HIV target nucleic acid is bound to a capture probe attached on the solid support, and detected by a biotinylated signaling probe to the captured target. This assay may also be suitable for the quantitative assessment of viral loads.

The PCR technique is remarkably efficient for the amplification of specific sequences of HIV proviral DNA. Also, it has been applied successfully for amplifying viral RNA by first transcribing the RNA to complimentary DNA with RT, and then amplifying the complementary DNA. However, this approach is less sensitive than using DNA directly as a template. The PCR technique has been used successfully for demonstrating nucleic acid of HIV-1 and -2, and HTLV-I and -II in a variety of specimens from infected individuals. The performance of the technique depends on the availability of sequences of the viral nucleic acid that flank a desired target sequence. The more commonly used sequences (primers) for HIV are derived from the *gag* and/or *pol* genes, which are the more conserved regions of the viral genome. Selection of sequences from the conserved regions makes it possible to detect viral variants both within and between infected individuals. Two primers are prepared using the flanking sequences, one complementary to the negative strand DNA and one complementary to the positive strand DNA. A thermostable polymerase is used to extend the primers in the presence of deoxynucleoside triphosphates. The cycle of DNA amplification is initiated by placing lysed specimens or extracted DNA in the thermocycler at a high temperature to denature the DNA. A relatively low temperature is then used to anneal the primers to the complementary region of the target DNA in the presence of a large molar excess of the primers. The primers hybridize to opposite strands of the target sequences and prime enzymatic extension along the nucleic acid template by exposure to an intermediate temperature. The end product is denatured by exposure again to high temperature, to begin another cycle. Each cycle of PCR results in a doubling of the target sequence. For example, after 20 cycles of amplification 1 million copies of the DNA segment are produced from a single copy of the target sequence. The amplified gene product is then detected by radio- or enzyme-labeled probes specific for the amplified sequence. An alternative method is to label the primers, which allows for direct detection of the product after capture by hybridization to an immobilized capture probe.

Careful choice of the HIV-positive and -negative controls for PCR is important, and should also include samples with no DNA. The PCR technique must be conducted with extreme care to avoid cross-contamination of samples or carryover of amplified products that can yield false-positive results. Precautions to avoid these possibilities include the use of positive displacement pipettes, preparing reagents and processing specimens in an area free of PCR-amplified products, sterilizing reagents, and the frequent change of gloves.

Although PCR has not been adapted for routine diagnosis of HIV, this technique has demonstrated clinical utility in several areas of research. These include:

1. The detection of HIV gene sequences during early infection, or prior to seroconversion
2. Resolution of the infection status of individuals with indeterminate serological results
3. Detection of HIV gene sequences in antibody positive individuals whose cells are negative by culture and/or antigen assays
4. Resolution of the infection status of babies born to seropositive mothers
5. Monitoring viral load in patients receiving therapeutic agents
6. Description of the genetic variability among HIV isolates

The PCR technique is highly sensitive and specific, but is subject to technical false-positives and negatives if not performed with the utmost of care. It is not a simple procedure, and requires careful and exact laboratory technique. At present, PCR is not suitable for use as a routine procedure, but it is an important research tool. However, recent developments in PCR technology suggest that the technique may soon be practical for the routine laboratory. This is based on a modification of the procedure in order to reduce the chances of obtaining false results due to contamination.

III. INTERPRETATION OF CONFIRMATORY TEST RESULTS

Most diagnostic serologic assays produce results that can be characterized as positive, negative, or as yielding a titer. In some cases, a result may be close to the cutoff between positive and negative (borderline reactor). Even the determination of titers can yield a result that may be on the border between significant and insignificant results. Therefore, it is sometimes difficult to determine the true meaning of some serological results.

The serologic diagnosis of HIV infection is an example of a situation in which the physician and the patient expect to obtain a conclusive result.

Unfortunately, HIV results are not always either positive or negative. There are many times that a firm conclusion cannot be determined. In these instances, the results are labeled as inconclusive or indeterminate.

Since ELISA results are read spectrophotometrically, the results usually do not present a problem, unless the O.D. value is close to the cutoff, in which case additional testing may be necessary. Although no consensus exists regarding the terminology for reporting of results, it is generally accepted that screening test results are reported as reactive or nonreactive, while confirmatory results are reported as positive or negative. It is recommended that the terminology stated in the test kit package insert be followed.

Confirmatory assays are usually read subjectively by the individual performing the test; the individual must decide whether the result is positive, negative, or indeterminate. In addition assays such as the WB result in profiles; thus many combinations of reactions are produced to identify and interpret. The interpretation of reaction intensities and overall patterns cannot be controlled objectively, and therefore variations in conditions and expertise among laboratories may cause further problems with results and interpretation. Tests such as the IFA require similar careful examination and interpretation to differentiate specific from nonspecific reactions, and to detect very weak reactions. Results of the IFA are also read subjectively, and may be dependent on the quality and optics of the fluorescence microscope.

A. Criteria for Positivity

Integral to every serological test are the guidelines or criteria that serve to determine whether the result fulfills the requirements to be labeled as positive, negative, or indeterminate. HIV testing has very stringent criteria, since results have such an important impact. These criteria are determined in two ways:

1. Manufacturers of test kits have predetermined the requirements for results based on studies of individuals who have been classified as positive or negative by other means (clinical status, culture, etc.). They then adjust their controls so that the test will identify the maximum number of positives, while not producing many false-positive results from individuals who are not infected.
2. Organizations having expertise in HIV testing can establish criteria based on large-scale evaluations. This is particularly true for the interpretation of WB results, discussed below. Unfortunately, these organizations do not always concur on their recommendations.

1. Western Blot Criteria

The classification of WB results is determined by criteria that have been established to yield the most accurate diagnosis. It is now universally accepted that a negative result is the absence of all bands on the blot. Two organizations, including the World Health Organization (WHO), suggest that results can also be reported as negative if there is only a very weak p17 band. Unfortunately, sera from some noninfected individuals will show some reactivity to one or more antigens. This may occur in as many as 15% of normal noninfected persons, and many times occurs in persons who are nonreactive by screening assays. Therefore, if ELISA nonreactive sera are tested by WB, many will result in an indeterminate profile. Most indeterminate results show only weak reactions to the gag proteins (mostly p17, p24, and/or p55); other patterns occur but are less frequent. Any WB reactivity that does not meet the requirements for being positive or negative (see below) must be considered as indeterminate. The significance of an indeterminate WB result is discussed under "Inconclusive Results".

No universally accepted criteria define a positive WB profile. Several organizations have formulated their own requirements for a WB result considered to be positive. The lack of consensus among these organizations reflects the still-incomplete understanding of the immune response, or disagreement in the interpretation of reactive bands. There is inconsistency in the immune responses from some individuals, making a general interpretation of results difficult. It is also difficult to determine whether certain antibody profiles are indicative of true infection. Some individuals who exhibit reactivity to p24 and p55 will later be shown to seroconvert, indicating that a p24 and p55 profile can be indicative of early infection. Conversely, other individuals may have the identical profile for long periods of time (years) and never exhibit evidence of seroconversion or even seroprogression (i.e., they are not infected). In fact, most indeterminate WB results from noninfected individuals exhibit the p24 and/or p55 profile.

Of some concern is the presence of oligomeric forms of the glycoproteins on WBs. It has been demonstrated that the gp160 on blots may actually be tetramers of gp41, occurring during the disruption and electrophoresis procedures (i.e., they may represent the same antigen). This could complicate the interpretation of positive results, especially if using criteria that include the presence of two envelope bands for a final result of positive. The degree to which this interference occurs or its effect on accurate diagnosis is not presently known.

Some kits licensed by the FDA use criteria that maximize the specificity of the assay for persons for whom little clinical or virologic information is available. These criteria may not be ideal for other situations in which patients have symptoms or are members of high-risk populations. Inevitably, criteria that confer a high specificity in the test will sacrifice some sensitivity, and vice versa.

In certain situations, such as diagnosis and the institution of therapy, a high specificity is needed so that only HIV-infected persons are treated. In other situations, such as for the detection of potentially infected blood donors, a high sensitivity is essential so that no infected persons are incorrectly identified as negative.

Supplemental Tests for HIV-1 Infection

Organizations such as the Association of State and Territorial Public Health Laboratory Directors (ASTPHLD) and the Centers for Disease Control (CDC) have chosen criteria that yield the highest combined sensitivity and specificity, i.e., the highest correct percentage of positive results and the lowest number of indeterminate results. The criteria used in some FDA licensed kits are the most specific, while those of ASTPHLD/CDC are the most widely used. Occasionally, a specimen may be considered positive by one set of criteria, while being classified as indeterminate by another. This should not be disturbing, since there are different reasons for adopting particular criteria. Each laboratory must choose the criteria that best meets specified objectives for diagnosis.

In the past, many different criteria existed for the interpretation of WB results; now there are essentially five. Table 3 provides a list of the different criteria currently in use and most widely adopted. Figure 32 shows a schematic representation of some WB results, and indicates the classification by the various organizations.

2. Indirect Immunofluorescent Assay Criteria

The IFA test requires some expertise to read, but does not present the interpretative problems noted with the WB. Most laboratories follow the same requirements for negative, positive, and indeterminate results:

A negative result is reported if:

- No cytoplasmic fluorescence at all is found in the infected or the noninfected control cells

A positive result is reported if:

- Cytoplasmic fluorescence is found at the initial dilution in the infected well, but not in the control well
- Cytoplasmic fluorescence is found at two twofold dilutions or greater in the infected well as compared to the control well

Note: All reactive wells must exhibit at least a 1+ intensity reaction.

An indeterminate result is reported if:

- The titer of reactivity is equal in the infected and the noninfected cells
- The fluorescence is in the nucleus
- The cytoplasmic fluorescence is only one, twofold dilution greater in the HIV-infected well than in the control well, or equal to or at a higher titer in the control well than the infected well

Indeterminate results usually indicate nonspecific reactivity, possibly due to the presence of autoantibodies.

TABLE 3
Criteria for the Interpretation of an HIV-1-Positive Western Blot

Organization	Criteria
ASTPHLD/CDC	Any two: p24, gp41, or gp120/gp160
FDA[a]	p24 and p31 and gp41 or gp120/160
American Red Cross	At least three bands, one from each gene product group, *gag*, *pol*, and *env*
Consortium for Retrovirus Serology Standardization (CRSS)	At least two bands: p24 or p31 and gp41 or gp120/160
World Health Organization (WHO)	At least two *env* bands

[a] Some FDA licensed WB kits.

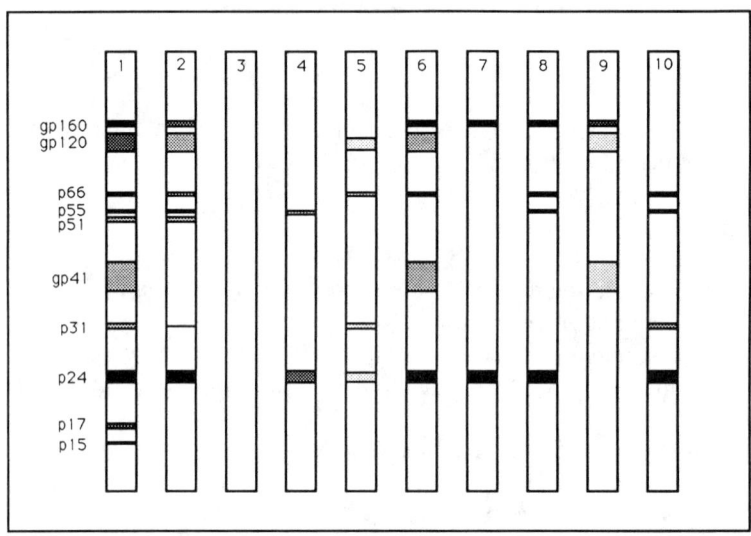

FIGURE 32. Examples of WB profiles and their classification according to recommended criteria Key: (1) strong positive control (positive by all criteria); (2) weak positive control; (3) negative control; (4) indeterminate (p24 and p55); (5) indeterminate (insufficient intensity of bands); (6) positive by all criteria, except indeterminate by FDA; (7) positive by ASTPHLD/CDC and CRSS criteria, indeterminate by others; (8) positive by ASTPHLD/CDC and CRSS and Red Cross, indeterminate by FDA and WHO; (9) positive by WHO and ASTPHLD/CDC, indeterminate by others; (10) indeterminate by all criteria, suggestive of possible HIV-2 infection.

B. Inconclusive Results

As previously stated, ELISA screening assays for HIV can almost always be reported as reactive or nonreactive since they are read objectively. Some screening tests such as the agglutination or dot-blot assays are read subjectively and may sometimes show an inconclusive result due to a reaction in the cell control. Confirmatory tests, however, can be positive, negative, or indeterminate, with an indeterminate classification being difficult to interpret. An individual with an indeterminate result should not be considered positive or negative until additional tests are performed, or until being monitored by retesting at a later time. Most authorities suggest that persons with indeterminate results should be retested after several months, although seroconversion may be detected in a shorter period of time. As stated in Chapter 8, the later retesting of an individual should be performed in parallel with the initial sample in order to ensure that the samples can be directly compared on the same run (with the same kit lot numbers and the same assay conditions).

WHO recommends that persons be tested after two weeks if highly suggestive WB profiles are produced. If an individual is retested over a period of six months and becomes negative or the band profiles do not progress to positive, HIV infection can generally be ruled out. For poorly understood reasons, many individuals continue to exhibit indeterminate results for years but are not infected. If an individual does progress serologically (seroprogression) or converts to positive (seroconversion) during retesting, the individual was probably infected at the time of the first test (early infection).

If a specimen exhibits sufficient reactions to antigens to be qualitatively positive, but the intensity of reactions is not strong enough to be considered positive (1+ or greater), the person is probably in a state of seroconversion and should be retested soon. However, similar patterns may indicate that a technical error occurred, such as carryover from adjacent wells.

The significance of an indeterminate WB result varies depending on the risk factors and the clinical status of the patient, and the WB profile produced. For example, individuals with a history of high risk behavior are more likely to be the ones who will later exhibit seroconversion, since the chances of their being infected are high. In addition, some WB profiles are much more suggestive of early infection (e.g., p24 and p55) than are others (e.g., p17 only). Many initially indeterminate results that subsequently become negative or remain indeterminate are probably a result of:

- Technical error (contamination or carryover from pipetting or splashing)
- Nonspecific reactions between antibodies to residual cell components of the infected cells; e.g., autoantibodies reacting to epitopes of the culture cells
- The presence of hypergammaglobulinemia, which may result in reactions to many antigens

- The presence of antibodies to some parasitic agents such as malaria
- Infection with an unknown, but related retrovirus

The occurrence of reactions to nonviral-specific antigens presents difficulty in interpretation. Although such reactions usually are not associated with infection, it is recommended that they be reported as inconclusive and a follow-up sample requested. Some organizations consider these reactions to be insignificant and report them as negative. Sometimes a WB strip will exhibit a high background; the whole strip will show some color, making interpretation difficult. This type of reaction should not be reported as negative, since proteins binding to the entire strip may mask reactions to specific antigens. If the results are repeatable with no change, use of another type of test should be considered, or a new specimen requested.

It is also known that some individuals with AIDS may lose reactivity to p24 and perhaps other antibodies later in disease, so that even AIDS patients may have indeterminate WB results by some criteria. Ancillary tests may be helpful in resolving these indeterminate results if the diagnosis is in question. The physician should use clinical data and the laboratory should use all available tests to determine the true status of the individual. In this case, if the original indeterminate result was produced by the WB, an IFA may subsequently be performed. The FDA has recently recommended that the IFA can be used to resolve an HIV-1 indeterminate result. For example, an IFA positive or negative result on such a specimen is evidence of infection or noninfection, respectively (see Figure 33). Other tests may be helpful, including the newer synthetic peptide tests, antigen tests, and perhaps specific IgM assays. The latter may be positive if the patient is truly in early infection. Additionally, tests such as the PCR, viral culture, and *in vitro* antibody production assays can be performed to help elicit the patient's true status. Also, some patients may be infected with HIV-2 and produce indeterminate results by HIV-1 assays; therefore, tests to detect HIV-2 infection should also be performed (Chapter 5).

Controversy exists over whether blood units from individuals who are WB indeterminate should be allowed to donate if their status has not changed after six months. The FDA, however, requires that a blood donor must test HIV negative by both the WB and the ELISA for six months in order to requalify for blood donation. These requirements are currently being revised by the FDA, and a recommendation is forthcoming.

IV. ADVANTAGES AND DISADVANTAGES OF CONFIRMATORY TESTS

Confirmatory tests to detect antibodies to HIV are generally more expensive and more time consuming to perform than screening tests. Also, most tests accepted as effective confirmatory assays are read subjectively, and therefore require a certain level of technical expertise (experience and knowledge). The major advantage of confirmatory assays is their excellent specificity. If the tests are performed correctly, and with no technical errors, there should be few false-positives. This specificity of the confirmatory tests is an absolute necessity, since the result of these tests will ultimately determine the status of the individual. In addition some confirmatory tests such as the WB can be used as a prognostic indicator. For example, profiles that show a decrease in anti-p24 over time may indicate disease progression. Another advantage of the use of additional tests is that the positive predictive value (Chapter 7) of a confirmatory or supplemental test is greatly increased because the prevalence rate of the sera being tested has increased, as only those sera reactive during screening are submitted for confirmatory testing.

Although the WB assay has the advantage of being able to indicate the actual specificities of antibodies that react with the HIV antigens, it is also the most expensive, time consuming, and the most difficult serologic test to interpret. In addition, it is an assay that is very technique dependent, since cross-contamination can easily occur. Despite the disadvantages of the WB, it does yield useful information and is considered by many as the best and most reliable confirmatory test for HIV infection.

The IFA test is also widely used by many testing centers and offers several advantages over the WB:

- The IFA is easier to perform, is less expensive than the WB, and can be completed in a relatively short time period.
- Unless autoantibodies are present in the sample, the test is generally easy to read.

The IFA does require the use of an expensive fluorescence microscope, some expertise to interpret the results, and more time to read the results if large numbers of specimens are tested at one time. Also, if a patient has received a therapy such as fluorescence angiograms, false-positive results may occur. The sensitivity and specificity of the IFA are equivalent to the WB. Similar to the WB, standardized kits are available from commercial manufacturers. When tests are developed "in house", the consistency of the reagents cannot be ensured, and results may differ between laboratories.

The use of certain combinations of screening tests can yield a high degree of specificity, and may be considered for use as confirmatory or supplemental tests. The use of two screening tests is usually less expensive than the use of a confirmatory test, and offers other advantages such as easier interpretation, ease of operation, and can be performed in testing centers (such as peripheral laboratories) where confirmatory tests are not available. In this confirmatory testing scenario, a second screening test with a different principle from the first screening test is used. For example, if an ELISA using a viral lysate is used for the screening test, a test incorporating a recombinant antigen or synthetic peptide antigen should be considered for supplemental testing. Similarly, a combination of indirect and competitive ELISAs may be considered. When results of both tests are reactive, the positive predictive value is very high. Usually, the noncompetitive (indirect ELISA) is used as the screening assay followed by a competitive assay that is considered to be more specific. If the first test is positive but the second is negative, additional confirmatory tests must be performed.

V. TESTING ALGORITHMS

The term "algorithm" refers to the sequence in which different tests are performed. Although each laboratory must determine its own testing algorithms based on the available tests, there are some general guidelines that should be followed for HIV testing. First, screening tests must be repeated in duplicate on all initially reactive samples. This is a measure to ensure that technical error did not occur; e.g., handling errors, testing of the incorrect sample, etc. The optimal situation would be to have the repeat tests performed on serum from the original specimen tube. For the sample to be considered as truly reactive, it must be reactive on at least two of the three tests. Second, all results that are repeatedly reactive on screening tests must be confirmed by a recognized confirmatory test or another established testing algorithm. This is to ensure that the result was not due to a biologic false-positive; e.g., due to nonspecific antibodies such as those that may occur in some autoimmune diseases.

The most commonly used testing algorithm is to screen sera using an ELISA and to confirm repeatedly reactive specimens using the WB (this must be in accordance with local regulations). This conventional strategy does have some technological, practical, and cost disadvantages:

- ELISAs may not be an efficient method for screening small numbers of specimens (perhaps a rapid test may be more cost efficient).
- ELISA may not be appropriate for many small, poorly equipped laboratories, or where automation cannot be supported.
- The WB assay is expensive and technically difficult to perform and interpret.
- Repeating the ELISA and performing the WB for confirmation often result in long delays for test reports.

Alternative testing algorithms to the conventional ELISA/WB algorithm have been evaluated by several laboratories and are currently being evaluated by organizations such as the WHO. Many laboratories have adopted these alternatives after careful evaluation. These alternatives include, but are not limited to:

1. Using the ELISA and an IFA

Supplemental Tests for HIV-1 Infection

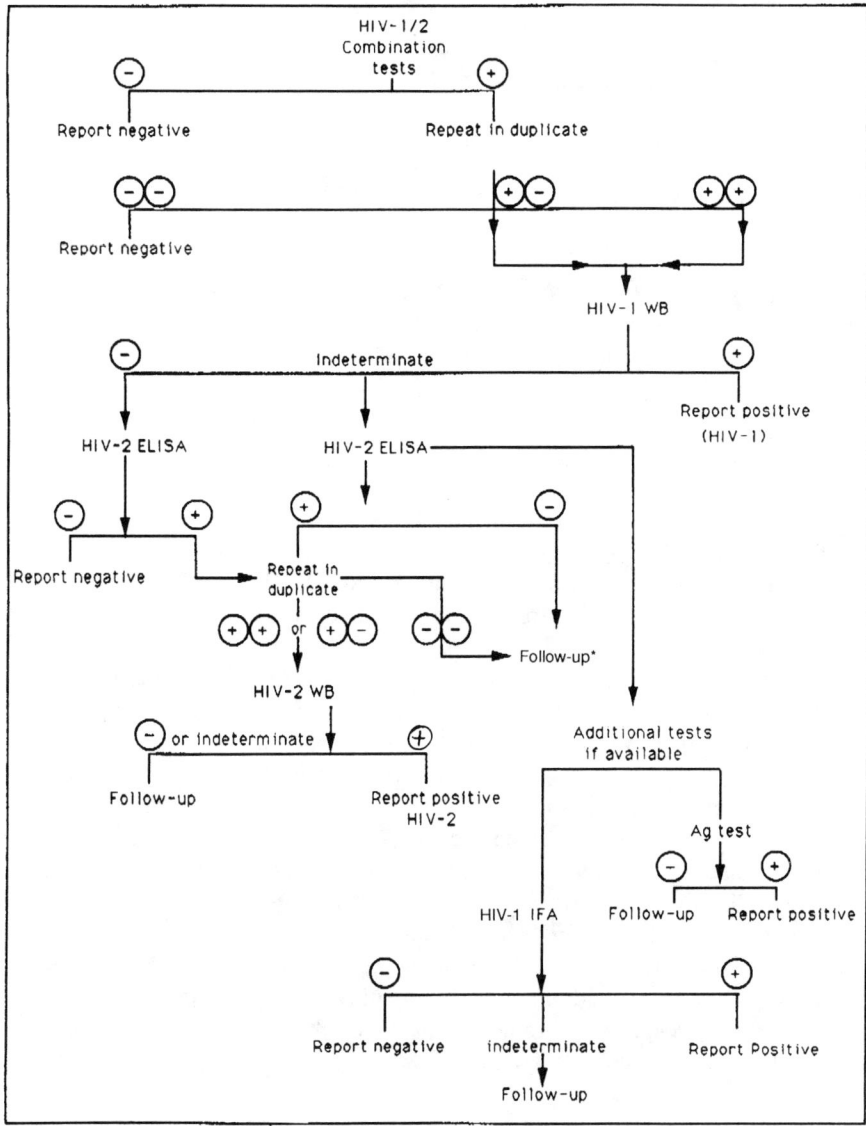

*If HIV-1 WB had been negative, can be reported as negative.

FIGURE 33. Testing algorithm for HIV-1 and -2 infection. Note that the specimens for follow-up should be reported as indeterminate.

2. Using two ELISA tests; e.g., competitive and noncompetitive, or a test using a viral lysate antigen combined with one using a recombinant or synthetic peptide antigens
3. Using a combination of a rapid or simple assay (dot-blot or agglutination test) with an ELISA

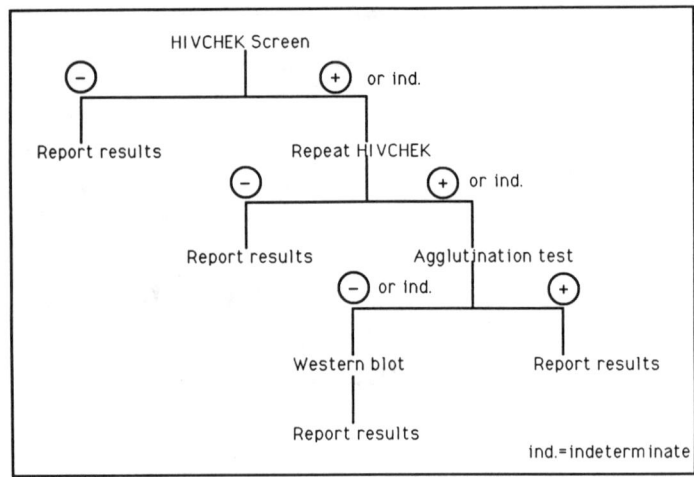

FIGURE 34. An example of an alternative testing algorithm.

4. Using a screening test in conjunction with an LIA
5. Use of a combination of two simple, rapid assays

Two examples of the many options for HIV testing strategies are presented below. Figure 33 indicates a proposed testing algorithm if identification of either HIV-1 or HIV-2 is sought, and the tests are available.

Figure 34 shows a testing algorithm based on the need for a less expensive testing strategy. This alternative strategy was evaluated by AIDSTECH/FHI and collaborators in several developing African nations including Kenya, Ghana, Senegal, and Zaire. It has been shown that a cost savings of up to 80% can be gained by using two rapid, simple tests in combination (as opposed to the conventional ELISA/WB), with minimum loss of sensitivity and specificity. However, it is critical that each laboratory or national program carefully evaluate a testing algorithm in their own setting before adopting it for routine use. Careful monitoring through quality control and proficiency testing is essential.

VI. COMMERCIALLY AVAILABLE CONFIRMATORY ASSAYS FOR HIV

Table 4 is a partial list of commercially available confirmatory or supplemental tests for HIV-1 and/or HIV-2, and indicates the principles of the tests and the nature of the antigens used. An extensive list of screening, rapid, and supplemental tests is noted in Appendix A.

TABLE 4
Supplemental Tests for HIV[a]

Test	Manufacturer	Type of Test	Antigen
Envacore	Abbott Laboratories	C	RP
Matrix HIV-1/2	Abbott Laboratories	Dot	S
Ancoscreen	Ancos	Blot	L
NovaPath[b]	BioRad	Blot	L
Speedscreen HIV-1/2	British Biotechnology	LIA	RP
HIV-1 Western blot[b]	Cambridge Biotech	Blot	L
HIV-2 Western blot	Cambridge Biotech	Blot	L
HIV-2 IFA	Cambridge Biotech	IFA	IC
Retro-tek IFA/HIV	Cellular Products	IFA	IC
RIBA HIV	Chiron	Blot	RP
PEPTI-LAV 2-2	Diagnostic Pasteur/Sanofi	LIA	SP
New LAV blot 1	Diagnostic Pasteur/Sanofi	Blot	L
New LAV blot 2	Diagnostic Pasteur/Sanofi	Blot	L
HIV-1 Western blot 1.1	Diagnostic Biotechnology	Blot	L
HIV-2 Western blot 1.2	Diagnostic Biotechnology	Blot	L
HIV-1/2 Blot 2.2	Diagnostic Biotechnology	Blot	L/SP
IgM Western blot	Diagnostic Biotechnology	Blot	L
HIV-1 Western blot[b]	DuPont deNemours/Ortho	Blot	L
Epiblot[b]	Epitope	Blot	L
Genetic Systems (HIV-1)	Genetic Systems/Sanofi	Blot	L
Genetic Systems (HIV-2)	Genetic Systems/Sanofi	Blot	L
INNO-LIA HIV-1/HIV-2	Innogenetics/MUREX	LIA	SP/RP
HIV Western blot[b]	Organon Teknika	Blot	L
Liatek HIV-1/2	Organon Teknika	LIA	SP/RP
Virgo IFA	Pharmacia	IFA	IC
Serofluor IFA	Virion	IFA	IC
Fluorognost HIV-1 IFA[b]	Waldheim Pharm/Thermascan	IFA	IC

Key: C = competitive ELISA, RP = recombinant protein, SP= synthetic peptide, IC= infected cells, L = lysate, Blot= Western blot, LIA = line immunoassay, and IFA = indirect fluorescent assay.

[a] May not include all available tests.
[b] FDA licensed.

REFERENCES

AIDS 89 Summary. A Practical Synopsis of the Fifth International Conference, Philadelphia Sciences Group, Philadelphia, 1989, 1.

Bagnarelli, P., Menzo, S., Manzin, A., Varaldo, P. E., Montroni, M., Giacca, M., and Clementi, M., Detection of human immunodeficiency virus type 1 transcripts in peripheral blood lymphocytes by the polymerase chain reaction, *J. Virol. Methods*, 32, 31, 1991.

Barr-Sinoussi, F., Cherman, J. C., and Rey, F., Isolation of a T-lymphotropic retrovirus from a patient at risk for acquired immune deficiency syndrome (AIDS), *Science*, 220, 868, 1983.

Bell, J., The polymerase chain reaction, *Immunol. Today*, 10, 351, 1989.

Brinchmann, J. E., Albert, J., and Vartdal, F., Few infected $CD4^+$ T cells but a high proportion of replication-competent provirus copies in asymptomatic human immunodeficiency virus Type 1 infection, *J. Virol.*, 65, 2019, 1991.

Brown, C., Kline, R., Atibu, L., Francis, H., Ryder, R., and Quinn, T. C., Prevalence of HIV-1 p24 antigenaemia in African and North American populations and correlation with clinical status, *AIDS*, 5, 89, 1991.

Burke, D. S., Laboratory diagnosis of human immunodeficiency virus infection, *Clin. Lab. Med.*, 9, 369, 1989.

Cabrian, K., Shriver, K., Goldstein, L., and Krieger, M., Human immuno-deficiency virus type 2: a review, *J. Clin. Immunoassay*, 11, 107, 1988.

Callahan, J. D., Constantine, N. T., and Watts, D. M., More support for the use of HIV-1/2 combination assays, *J. Virol. Methods*, 28, 217, 1990.

Centers for Disease Control, Interpretation and use of the Western blot assay for serodiagnosis of human immunodeficiency virus type 1 infections, *MMWR*, 38, 1, 1989.

Committee on Human Retrovirus Testing, Fifth Consensus Conference on Testing for Human Retroviruses: Report and Recommendations, Kansas City, MO, 1990, 5.

Desai, S., Bates, H., and Michalski, F. J., Detection of antibody to HIV-1 in urine, *Lancet*, 337, 183, 1991.

Dodd, R. Y. and Fang, C. T., Eds., Confirmatory and Supplementary Testing, Internal report, Ortho Diagnostics Systems, 4, 1, 1990.

Fenovillet, E., Blanes, N., and Gluckman, J. C., gp160 of commercial HIV Western blots is not gp160 env: should criteria for seropositivity be revised?, *AIDS*, 5, 770, 1991.

Jackson, J. B., Human immunodeficiency virus type 1 antigen and culture assays, *Arch. Pathol. Lab. Med.*, 114, 249, 1990.

Jehuda-Cohen T., Slade, B. A., and Powell, J. D., Polyclonal B-cell activation reveals antibodies against human immunodeficiency virus type 1 (HIV) in HIV-1 seronegative individuals, *Proc. Natl. Acad. Sci. U.S.A.*, 87, 3972, 1990.

Keller, G. H., Huang, D. P., and Manak, M. M., Detection of human immunodeficiency virus type 1 DNA by polymerase chain reaction amplification and capture hybridization of microtiter wells, *J. Clin. Microbiol.*, 29, 638, 1991.

Kwok, S. and Higuchi, R., Avoiding false positives for PCR, *Nature*, 339, 237, 1989.

Mitchell, S. and Mboup, S., HIV testing, in *The Handbook for AIDS Prevention in Africa*, Family Health International, Durham, NC, 1990, 20.

Centers for Disease Control, Interpretation and use of the Western blot assay for serodiagnosis of human immunodeficiency virus type 1 infections, *MMWR*, 38, 1, 1989.

Ou, C. Y., Kwok, S., and Mitchell, S. W., DNA amplification for direction of HIV-1 in DNA of peripheral blood nuclear cells, *Science*, 239, 295, 1988.

Phair, J. P. and Wolinsky, S., Diagnosis of infection with the human immunodeficiency virus, *J. Infect. Dis.*, 159, 320, 1989.

Product Preview, New ENZO assay for detection of HIV DNA, *MT Today*, December 9, p. 10, 1991.

Report, PCR-based method determines circulating HIV levels, *Clin. Lab. Lett.*, 13, 4, 1992.

Sandstrom, G., Schooley, T. R., and Ho, D. D., Detection of human anti-HTLV-III antibodies by indirect immunofluorescence using fixed cells, *Transfusion*, 25, 308, 1985.

Schmidt, B. L., A rapid chemiluminescence detection method for PCR-amplified HIV-1 DNA, *J. Virol. Methods*, 32, 233, 1991.

Schochetman, G., Ou, C. Y., and Jones, W. K., Polymerase chain reaction, *J. Infect. Dis.* 158, 1154, 1988.

Sloand, E. M., Pitt, E., Chiarello, R. J., and Nemo, G. J., HIV testing, state of the art, *JAMA*, 266, 2861, 1991.

Sninsky, J. J. and Kwok, S., Detection of human immunodeficiency viruses by the polymerase chain reaction, *Arch. Pathol. Lab. Med.*, 114, 259, 1990.

Tsang, V. C. W., Peralta, J. M., and Simmons, A. R., Enzyme linked immunoelectrotransfer blot techniques (EITB) for studying the specificities of antigens and antibodies separated by gel

Chapter 5

HIV-2

I. INTRODUCTION

In 1985, several years after the recognition of HIV as an infectious agent, a similar virus that caused AIDS was discovered in West Africa. This led to the classification of the human immunodeficiency virus into two types: HIV-1 (the initial virus) and HIV-2. HIV-2 seemed endemic primarily in areas of West Africa, in contrast to the worldwide prevalence of HIV-1. However, the prevalence and distribution of HIV-2 infections have increased, and infections have now been described in many other areas including Europe, South and North America, the Persian Gulf Region, and in North and East Africa. Although most cases are linked to West Africa, it will probably not be long before cases of endemic transmission are described in many countries throughout the world. The testing of historical serum specimens has indicated that HIV-2 had been present in West Africa for several decades. In some countries, the prevalence of HIV-2 is greater than HIV-1, but HIV-2 infects the same risk groups and is transmitted in the same manner as HIV-1. The transmission of HIV-2 may be less efficient, and age-specific seroprevalence data indicate that HIV-2 rates of infection increase with increasing age, different than the age-specific seroprevalence of HIV-1. In the U.S., nearly 40 cases of HIV-2 have been verified, most of which have been linked to West Africa. One report in the U.S. described an HIV-2-infected individual who had hundreds of sexual contacts in the U.S., exemplifying the potential for enormous spread.

Sequencing of a limited number of HIV-2 isolates has revealed the same range of genetic variation for HIV-2 as for HIV-1, although HIV-2 isolates share genetic homology to HIV-1 of about 30 to 60%. By PCR analysis, HIV-2-infected individuals exhibit lower levels of proviral DNA in circulating lymphocytes. This lower level of viral load may be related to the lower efficiency of CD4 receptor binding and may be an explanation for the seemingly slower transmission and longer incubation periods observed for HIV-2 as compared to HIV-1. The efficiency of viral isolation from samples appears to be lower for HIV-2 as compared to HIV-1.

Biologically, HIV-2 is very similar to HIV-1. Overall, the HIV-1 and -2 genomes exhibit about 60% homology in the more conserved *gag* and *pol* genes, and 30 to 40% homology in the other viral genes and LTR sequences. The HIV-2 virus causes the same pathological consequences as HIV-1. HIV-2 may have a longer latent period, although this was recently challenged. The immune responses to HIV-2, the method of infection, and the pathology caused by HIV-2 are similar to those of HIV-1, and therefore are not described here (see Chapters 1 and 2).

Diagnostically, HIV-2 infections can present problems. In fact, the first two cases of HIV-2 infection were diagnosed in Europe when two West African patients were diagnosed clinically with AIDS, but were both negative by HIV-1 screening assays and indeterminate by the HIV-1 WB. The ability to detect antibody to HIV-2 is becoming increasingly important, as suggested by studies in some areas of Europe in which HIV-2 infection is regarded as a serious problem. In some cases, HIV-2-infected individuals had never had West African contacts. Tests designed to detect infection by HIV-1 do not always detect infection by HIV-2 and vice versa. Therefore, an extra measure of vigilance must be practiced in order to identify infections that might not be readily apparent using some HIV-1 assays. With the development of newer technologies, the two viruses may be differentiated serologically and by utilizing sophisticated techniques such as PCR.

II. ANTIGENS OF HIV-2

Similar to the composition of HIV-1, HIV-2 contains genes that code for proteins, glycoproteins, and enzymes; one also finds structural components and regulatory components that function in the same capacity as those of HIV-1. In general the structural antigens of HIV-2 differ slightly in molecular weight and in their amino acid composition. However, some variations exist in the structural antigens of HIV-2. For example, the transmembrane antigen shows polymorphism in the size of the molecule (mol wt = 32,000 to 41,000) probably due to differences in glycosylation. Within the same individual, the transmembrane antigen may have a mol wt of 32,000 when isolated from the serum, but a mol wt of 40,000 when isolated from brain tissue.

The structural genes code for proteins and glycoproteins corresponding to the groups and molecular weights indicated below:

gag
    p56   Precursor
    p26   Core
    p16   Matrix

pol
    p68   Reverse transcriptase
    p53   Reverse transcriptase
    p31/34   Endonuclease

env
    gp140   Precursor
    gp105/125   External
    gp 36/41   Transmembrane

The major regulatory genes *tat* and *rev* appear analogous in HIV-1 and HIV-2, although they may vary more in the HIV-2 virus. However, one of the regulatory components, vpx (p16), is present in HIV-2 but not in HIV-1; it is similar in function to the vpu of HIV-1, may be responsible for determining which cells are infected with the virus, and may be responsible for the different pathologic effects noted with HIV-2 infection. Antibody responses to vpx have been detected in about 14% of HIV-2-infected individuals. This may have important diagnostic application, since the detection of antibodies to vpx will identify HIV-2 infection, especially when sera react by both assays. Antibodies to the nef regulatory protein, which may not function as a down regulator in HIV-2, have been detected in about 25% of HIV-2-infected individuals. Sera that contain HIV-1 antibodies have no neutralizing affect on HIV-2 isolates, which contrasts with the cross-neutralizing activity of HIV-2 sera on HIV-1 isolates.

Antibodies to HIV-1 will frequently cross-react with the antigens of HIV-2 and produce positive reactions in serological assays designed to detect antibodies to HIV-2. Conversely, antibodies to HIV-2 may react in HIV-1 assays. However, in many cases HIV-1 serologic assays will not detect HIV-2 infections and HIV-2 tests will not detect HIV-1 infections. In a study conducted at the CDC, FDA-licensed HIV-1 EIAs were investigated for their ability to detect antibodies to HIV-2. The detection of HIV-2 antibodies ranged from 60 to 91% depending on which manufacturer's test was used. Other studies have indicated even lower rates of detection, with one indicating detection rates of 8 to 62%. Most cross-reactions represent antibody induced by the core and/or pol antigens, as these are highly conserved between the two different viruses.

The antigens of HIV-2 are similar to those of HIV-1, but the molecular weights may vary slightly. For example: the gag proteins of HIV-2 have molecular weights of p56 (precursor), p26, and p16; the pol proteins p68 and p34; and the envelope glycoproteins gp140 (precursor), gp105, and the transmembrane gp36 (or gp41). Some tests incorporated strains of HIV-2 (NIHZ) that contained envelope antigens of molecular weights: gp160, gp120, and gp41, identical to those of HIV-1, but antigenically distinct. Therefore, it is not uncommon to see the major core antigen referred to as p24 or p26, and the envelope antigens as gp160 or gp140, gp120 or 105/125, and gp41 or 36. Fortunately, most manufacturers of test kits have recently used strains of HIV-2 (ROD) and developed their methods to produce some consistency in antigen classification (as indicated in the listing above). Table 5 compares the antigens of HIV-1 and HIV-2 as they are most commonly classified now.

III. TESTING FOR HIV-2

A. Screening Tests

Because HIV-2 infections result in antibody responses similar to those of HIV-1 infections, there are usually immune responses to all of the major gene

TABLE 5
Comparison of the Antigens of HIV-1 and HIV-2

Gene	HIV-1	HIV-2
env		
Precursor	gp160	gp140
External	gp120	gp105 (125)
Transmembrane	gp41	gp36 (41)
gag		
Precursor	p55	p56
Core	p24	p26
Matrix	p17	p16
	p9	
	p7	
pol		
RT	p66	p68
RT	p51	p53
Endonuclease	p31	p34

products. Although the actual molecular weights of the proteins and glycoproteins differ slightly, both viruses have very similar structure and chemical composition. It is not unusual to expect that cross-reactions will occur. In a number of HIV-2 positive sera, however, the results are negative or indeterminate by HIV-1 tests. By HIV-1 ELISAs, the O.D. readings of HIV-2 positive specimens may be high negative; by WB the results may be indeterminate. Therefore, it is important to recognize slightly high negative readings and suggestive indeterminate results by HIV-1 tests, and consider testing the serum using HIV-2 tests. When using HIV-2 tests, similar cross-reactions may be observed when testing HIV-1-positive sera.

The most common antibodies that cross-react between these two viruses are those directed toward the core proteins (e.g., p24/26 and p55). Similarly, antibodies to the pol antigens may also produce cross-reactions. The env antigens are more unrelated and generally will not produce cross-reacting antibodies. If a test primarily incorporates env antigens from a particular virus, it will be more specific for detecting antibodies to that virus. Conversely, tests using primarily gag antigens will detect both infections more readily, but not always. This may or may not be advantageous, depending on whether sensitivity or specificity is desired.

An HIV-2 ELISA has been licensed by the FDA since 1990. Screening tests for the detection of antibodies to HIV-2 are identical to those for HIV-1, with one exception — the antigens that are used. To adequately detect antibodies to HIV-2, antigens from the HIV-2 virus must be used. Although many antigens are shared between HIV-1 and HIV-2, each virus also contains specific antigens, and it is the antibodies to these specific antigens that must be detected in order to identify the specific viral type. With the exception of the antigen (HIV-2) that is coated on the solid phase, HIV-2 screening tests are identical to the HIV-1 tests, and therefore are not discussed here (see Chapter 3). Table 6 lists

TABLE 6
Some Commercially Available HIV-2 Screening Tests[a]

Test	Manufacturer	Type of Test	Antigen
ELISAs			
For the Detection of Antibody to HIV-2			
HIV-2 AB	Clonatec	I	SP
ELAVIA II	Diagnostics Pasteur/Sanofi	I	L
HIV-2 ELISA	Diagnostic Biotechnology	I	L
Select-HIV	IAF Biochem/Coulter	I	SP
HIV-2 EIA[b]	Genetic Systems/Sanofi	I	L
For the Detection of Antibody to HIV-1 and HIV-2			
IMx	Abbott Laboratories	F	RP
Rec. HIV-1/HIV-2 3rd generation[b]	Abbott Laboratories	S	RP
Enzygnost Anti-HIV-1 & 2	Behringwerke	I	SP
Biotest Anti-HIV-1/2	Biotest Diagnostics	I	RP
HIV-1/HIV-2 Modul-Test	Biochrom	I	SP
Peptide HIV ELISA	Cal-Tech Diagnostics	I	SP
HIV (1 + 2)	Clonatec	I	SP
ELAVIA mixT	Diagnostics Pasteur/Sanofi	I	L
Rapid ELAVIA mixT	Diagnostics Pasteur/Sanofi	I	L
Genelavia mixT	Diagnostics Pasteur/Sanofi	I	SP/RP
DuPont HIV-1/HIV-2 ELISA	DuPont de Nemours	I	SP/RP
HIV-1/HIV-2 EIA[b]	Genetic Systems/Sanofi	I	L
Anti-HIV-1/HIV-2 EIA	Hoffmann-LaRoche	I	RP
Human HIV 1 & 2	Human	I	RP
Detect-HIV-1/-2	IAF Biochem/Coulter	I	SP
Select HIV-1/2 Diff	IAF Biochem/Coulter	I	SP
Innotest HIV-1/-2	Innogenetics	I	SP/RP
Vironostika HIV-1 + 2	Organon Teknika	I	L/SP
Wellcozyme HIV-1 + 2	Wellcome Diagnostics	S	SP/RP
GACELISA	Wellcome Diagnostics	S	RP
For the Detection of Antibodies to HIV-1, HIV-2, HTLV-I, and HTLV-II			
Bioelisa HIV-1 & 2, HTLV-I & II	Biotest Ltd.	I	SP
Detect Plus	IAF Biochem International	I	SP
Rapid and Simple Assays			
For the Detection of Antibody to HIV-1 and HIV-2			
Test Pack HIV-1/HIV-2 AB	Abbott Laboratories	Dot	RP
Rap. Test Dev. HIV-1/2 Recomb.	Cambridge Biotech	Dot	RP

TABLE 6 (Continued)
Some Commercially Available HIV-2 Screening Tests

		Type of	
Test	Manufacturer	Test	Antigen

For the Detection of Antibody to HIV-1 and HIV-2

Rapid HIV-1/2	Clonatec	Dot	SP
HIV-SPOT	Diagnostic Biotechnology	Dot	SP/RP
HIVCHEK 1 + 2	DuPont de Nemours	Dot	SP/RP
Genie HIV-1/-2	Genetic Systems/Sanofi	Dot	SP
SUDS 1 + 2	MUREX	Dot	L/SP
Immunocomb Bi-Spot	PBS Organics	Dot	SP
Recodot	Waldheim Pharmazeutika	Dot	RP

Key: L = lysate, RP = recombinant protein, SP = synthetic peptide, LIA = line immunoassay, CAP = capture assay, A = agglutination assay, Dot = dot-blot or microparticle assay, SP= synthetic peptide, I = indirect ELISA, and S = sandwich ELISA F = fluorescence.

[a] May not include all available tests.
[b] FDA licensed.

some commercially available HIV-2 screening tests, confirmatory tests for HIV-2 can be noted in Table 4 and in Appendix A.

Recently, tests have been developed using antigens from both viruses in order to detect either infection (see Section III.C). The differentiation of HIV-1 and -2 infections can usually be accomplished using highly specific ELISAs (e.g., synthetic peptide-based), WBs, RIPAs, or the PCR (Chapters 3 and 4). The use of recombinant antigens corresponding to three viral proteins, HIV-1 core and envelope and HIV-2 envelope, allows for improved detection of HIV-1- and HIV-2-positive specimens simultaneously.

B. Confirmatory Tests

Confirmatory tests such as RIPA and WB are infused with the ability to differentiate HIV-1 and HIV-2 infections since the molecular weights of the antigens differ. RIPA (Chapter 4) is considered by many to be the most reliable method for confirming HIV-2 infection, although it is also one of the most laborious. Recently, EIA tests using chemically synthesized peptides corresponding to unique immunogenic regions within the respective transmembrane glycoproteins have been shown to exhibit good correlation with WB and RIPA for identifying and differentiating HIV-1 and HIV-2 antibodies. Furthermore, these tests are valuable for differentiating samples that produce reactions to both viruses (dual reactors).

HIV-2 WBs are available from several manufacturers, although the strains used differ and produce blots exhibiting different molecular weights, as mentioned above. Similar to the HIV-1 WBs, HIV-2 blots use partially purified and

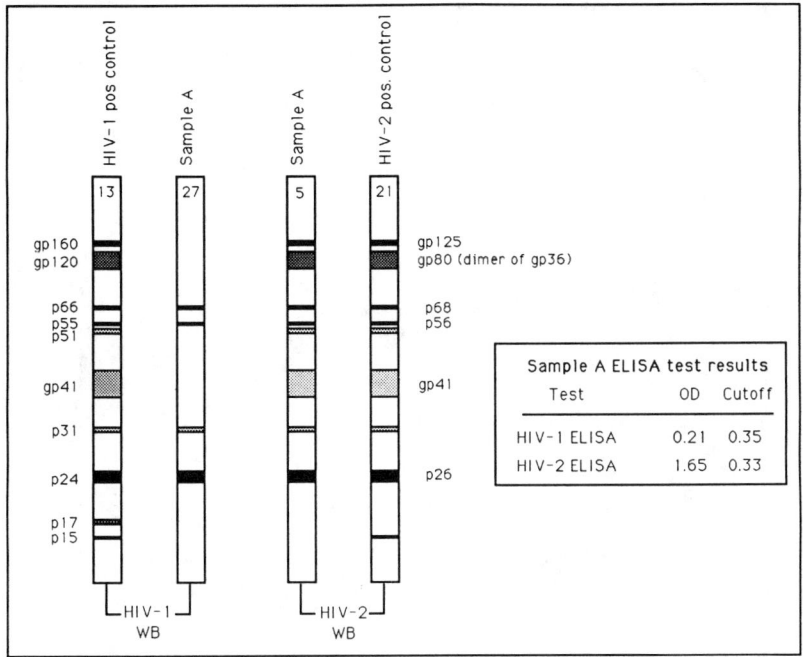

FIGURE 35. A typical reaction of an HIV-2-positive sample by HIV-1 and HIV-2 WBs and ELISAs.

inactivated viruses that have been grown in human cell lines such as HUT-78 cells. The tests are similar in principle and performance to those of HIV-1 (see Chapter 4).

As discussed previously, samples that contain antibodies to HIV-2 may yield indeterminate results when tested by an HIV-1 WB. Although some HIV-2 -positive samples will react with all antigens on the HIV-1 blot, it is not uncommon to have cross-reactions only with the gag or the gag and pol antigens. In fact, the general rule is that cross-reactions will not occur with the envelope antigens, at least not to a significant level. Therefore, when an HIV-1 WB profile exhibits reactions to gag and/or pol, without reactivity to env, HIV-2 infection should be considered, and an HIV-2 ELISA and/or WB performed. A typical reaction of an HIV-2-positive sample using HIV-1 and -2 ELISA and WBs is shown in Figure 35.

C. HIV-1/HIV-2 Combination Assays
1. Combination Screening Tests

In late 1991, the FDA licensed the first combination HIV-1/HIV-2 test and recommended that blood banks start screening for HIV-2 by mid-1992. Blood banks in the U.S. will be able to use either the previously licensed HIV-2 ELISA screening test and the HIV-1 ELISA, or one of the two licensed HIV-

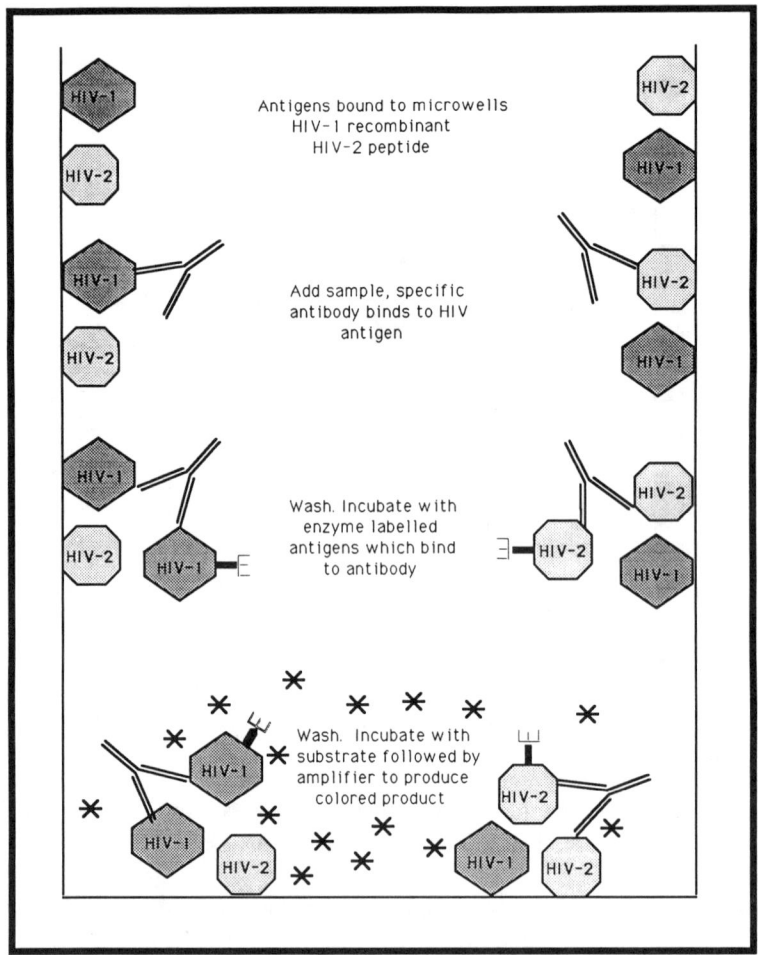

FIGURE 36. A combination HIV-1/HIV-2 antigen sandwich ELISA (Wellcozyme).

1/-2 combination tests. Samples that test positive by the combination test would be tested by an HIV-1 WB. If negative or indeterminate by this test, a specific HIV-2 test would be used to confirm the positive result, since there is currently no licensed HIV-2 WB or other licensed confirmatory test (see Figure 33).

Most second generation combination screening tests use the indirect format and do not identify which antibodies are present (anti-HIV-1 or -2); a positive reaction only indicates that antibodies to one or the other are present. Recently, third generation combination assays have been developed in both microtiter and bead formats. These assays utilize recombinant and/or synthetic peptide antigens to capture antibodies to either HIV-1 or -2, with labeled HIV-1 and

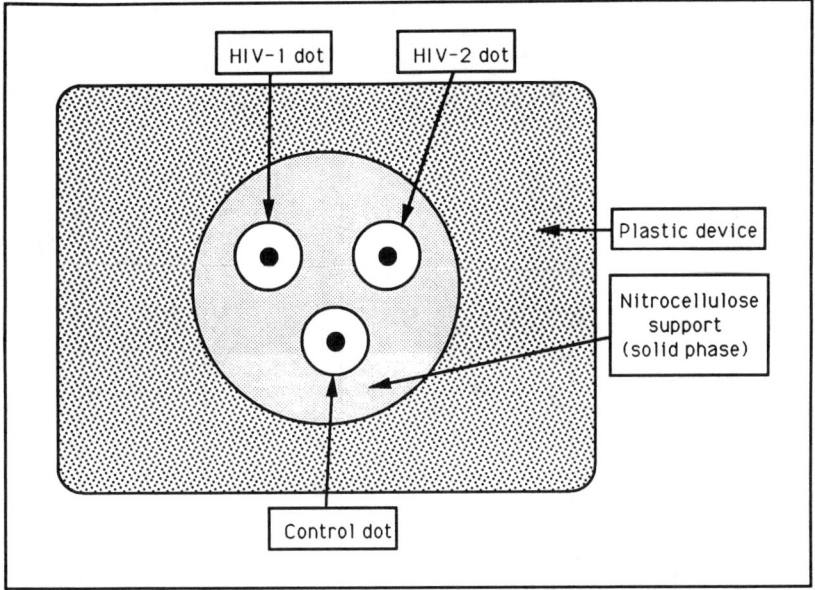

FIGURE 37. A combination HIV-1/HIV-2 dot-blot assay to differentiate infection.

-2 antigens subsequently being incorporated to bind to the antibodies (Figure 36). As discussed in Chapter 3, these antigen sandwich tests have the advantage of being able to detect both IgM and IgG antibodies simultaneously, and therefore may offer an increased sensitivity. Some dot-blot combination assays are available that have separate dots containing antigens for each virus. With these tests, differentiation of HIV-1 and HIV-2 infections can be accomplished with one test (Figure 37). This saves time and money. Combination tests are more expensive than single detection tests, but are less expensive than using separate screening tests for HIV-1 and HIV-2. Tests currently under evaluation can detect HIV and hepatitis B simultaneously, and one is available that is capable of detecting antibodies to HIV-1, HIV-2, HTLV-I, and HTLV-II simultaneously (see Appendix A).

2. Combination Confirmatory Tests

WB assays have been developed that have the ability to identify and differentiate infections by HIV-1 and -2. Most incorporate the use of viral lysates from HIV-1 and synthetic peptides artificially applied from HIV-2 on the same nitrocellulose strip (a modified or augmented WB). In this case, multiple HIV-1 antigens and one HIV-2 specific band (gp36 or gp41) are present on the strip; antibodies to the HIV-2 antigen are considered to be diagnostic if a reaction is present. These augmented blots usually also contain a control line, in an effort to indicate that the test has been performed properly. Figure 38 depicts a typical combination (augmented) WB assay. Each com-

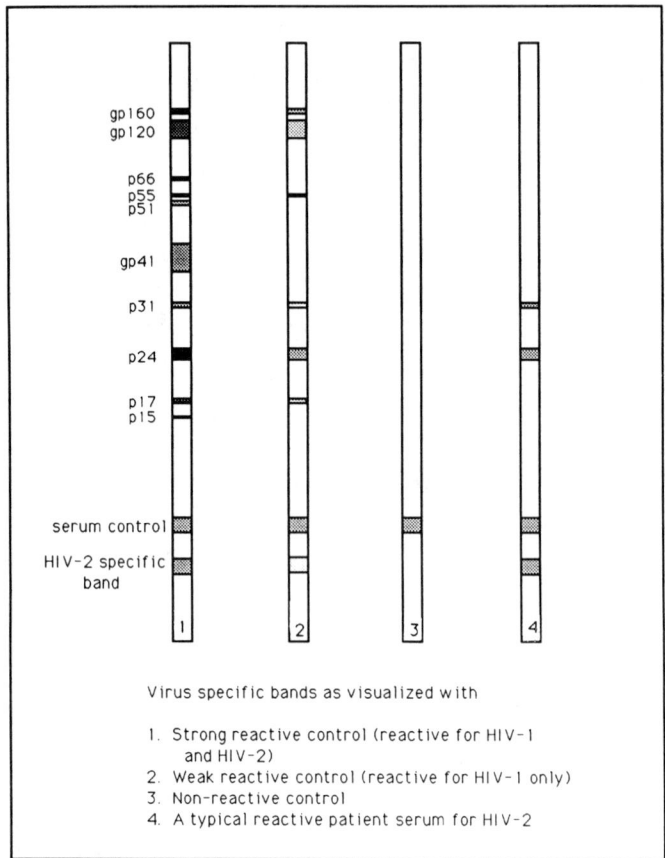

FIGURE 38. An example of a combination WB (augmented) assay.

pany that markets these tests indicates in their package inserts the criteria required for positivity. The criteria established by Diagnostic Biotechnology include reactions to one gene product from each of the three major groups (*gag*, *pol*, and *env*) for positivity for HIV-1. To be considered positive for HIV-2, reactions to the HIV-2 specific antigen plus a reaction to HIV-1-specific antigens (but which do not meet the criteria for positivity for HIV-1) must be present.

Recently, combination tests have become available that contain artificially applied antigens from both HIV-1 and HIV-2 in a WB format. The LIAs have gained in popularity and are based on the application of recombinant and synthetic peptide antigens on a plastic strip (Chapter 4). One particular combination assay includes three recombinant HIV-1 proteins (p24, p17, and p31) and two synthetic peptide antigens, one corresponding to the envelope antigens of HIV-1 (gp41) and one to the envelope antigens of HIV-2 (gp36). The use

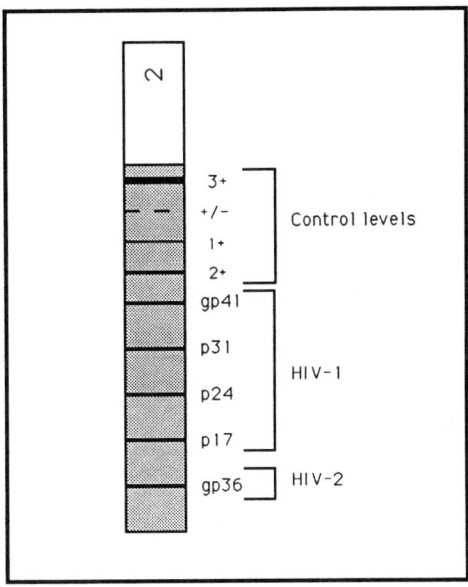

FIGURE 39. A combination LIA.

of these antigens helps to eliminate the problems of indeterminate and atypical results that may occur due to antibodies to cellular components in viral lysate-based confirmatory assays. In addition the problem of insufficient quantity of certain viral antigens, from virion preparations, which can lead to an underestimation of certain antibodies, can be circumvented by the mechanical application of antigen. The LIA is easy to perform, requires a total time of 2.5 h, and produces good performance with filter paper and saliva samples. An example of the LIA is presented in Figure 39.

Another combination confirmatory test has been introduced by Abbott Laboratories. The MATRIX is a semiautomated system that utilizes a dot-blot principle (Chapter 4). The MATRIX HIV-1/HIV-2 uses highly purified recombinant proteins on a nitrocellulose-based solid phase to capture specific HIV antibodies in the sample. The system employs a test cell in which a number of HIV-1 and HIV-2 antigens have been applied as distinct spots on a nitrocellulose membrane. The captured antibodies are detected by means of an enzyme-substrate indicator system, and are read by the optics contained within the instrument system. Therefore, antibody profiles are produced, similar to the WB assay. This test has the capacity to test a large number of samples in a relatively short period of time and with minimum hands-on time. The system introduces its own criteria for positivity, using the ASTPHLD/CDC criteria (Chapter 4) for positivity for HIV-1, and for HIV-2: reactions to gp41 of HIV-2 plus reactivity to any one of p17, p24, p31, or p66 of HIV-1. If the criteria to both HIV-1 and HIV-2 are met, then the specific infection is differentiated by comparing the ratios of reactions to the transmembrane antigens of HIV-1 vs. HIV-2.

Combination assays have been used with much success and have detected HIV-2-positive sera not identified using HIV-1 screening tests. There is no loss in sensitivity of these assays for detecting HIV-1 antibodies when compared to tests designed to detect antibodies to HIV-1 only.

At least five commercial companies are marketing HIV-1/-2 combination assays. As with HIV-1 and -2 tests, a variety of formats are available, including beads, microtiter wells, and dot-blots. There are similarities and differences between the combination assays:

- Most incorporate recombinant and/or synthetic peptide antigens in order to obtain adequate specificity
- Some use a combination of two synthetic peptides and others as many as four
- Some use both recombinant and synthetic peptides simultaneously, or viral lysates and synthetic peptides simultaneously

D. Interpretation of Test Results

Defining criteria to describe the characterization of HIV-2 results by confirmatory assays has fostered much reluctance among researchers. This is due in part to the lack of understanding of the exact immune responses to HIV-2, and to the lack of large numbers of seroconversion samples to investigate. Furthermore, no current standardization exists in the preparation of HIV-2 WBs, and therefore, interpretation of the reactivity to the various antigens is difficult. Furthermore, individuals may be infected with different strains of HIV-2, resulting in varying reactivity to different antigens by the different manufacturers' WBs.

Most organizations that have created criteria agree on the criteria for a negative reaction; i.e., the absence of all bands to HIV-2 viral-specific antigens. For a positive classification, WHO requires reactivity to at least two HIV-2 envelope antigens. Other organizations have suggested that a positive result is indicated by reactivity to p26 (gag), and gp34 or gp105 (env). Unfortunately, a consensus for positive criteria has not been established. At minimum, reactivity to an envelope antigen of HIV-2 must be present to even suggest HIV-2 infection. An indeterminate profile is one in which the criteria for a positive and a negative profile are not met. The use of these criteria are claimed to yield maximum specificity in order to differentiate between HIV-1 and -2. The same recommendations noted in Chapter 4, Section III are applied to HIV-2 indeterminates; i.e., follow-up testing. Samples that exhibit significant reactivity to antigens of both viruses should be further resolved using more sophisticated technologies (synthetic peptides or PCR). If a specimen is tested by both HIV-1 and HIV-2 WBs, the blot exhibiting the strongest reactivity to envelope antigens usually indicates which infection is present. Figure 40 shows a sche-

HIV-2 101

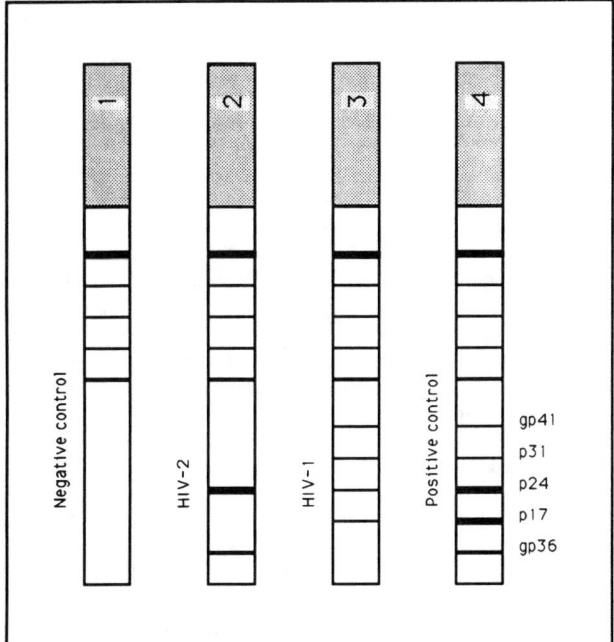

FIGURE 40. Examples of reactions using a combination LIA (Organon).

matic representation of some possible results obtained using a combination LIA.

It is the policy in many laboratories, and mandated in some areas, to test all HIV-1 indeterminate sera using HIV-2 assays, unless only minor reactions occur to a few antigens (e.g., p17 or p55). Any HIV-1 indeterminate result exhibiting reactions to the *gag* and *pol* antigens, but not to the *env* antigens, is highly suggestive of HIV-2 infection (see Chapter 4, Section III).

REFERENCES

Adams, C. W., Beatrice, S., Cabrian, K., Krieger, M., Majewski, S., and Shriver, M. K., Patterns of serological cross reactivity of HIV-2 antibody-positive specimens on HIV-1 Western blot, in Program and Abstracts of the 6th Int. AIDS Conf., Abstract FA207, San Francisco, June 1990, 125.

Benito-Garcia, A., Goncalves, H., and Pista, A., HIV-2 in Portugal: situation in 1990, 6th Int. Conf. AIDS, Abstract FC662, San Francisco, June 1990, 662.

Busch, M. P., Petersen, L. R., Scable, C., and Perkins, H. A., Monitoring blood donors for HIV-2 infection by testing anti-HIV-1 reactive sera, *Transfusion*, 30, 184, 1990.

Cabrian, K., Shriver, K., Goldstein, L., and Krieger, M., Human immunodeficiency virus type 2: a review, *J. Clin. Immunoassay*, 11, 107, 1988.

Callahan, J. D., Constantine, N. T., and Watts, D. M., More support for the use of combination assays, *J. Virol. Methods,* 28, 217, 1990.

Centers for Disease Control, MMWR surveillance for HIV-2 infection in blood donors—United States, *MMWR,* 39, 829, 1990.

Clavel, F., Guetard, D., and Brun-Vezinet, F., Isolation of a new human retrovirus from West African patients with AIDS, *Science,* 288, 343, 1986.

Constantine, N. T., Callahan, J. D., and Watts, D. M., A time for HIV-1/HIV-2 combination tests?, *J. Virol. Methods,* 26, 219, 1989.

Constantine, N. T., Sheba, F., Fox, E., Woody, J. N., and Abbatte, A. E., Serological evidence for HIV-2 in East Africa, *Int. J. STD AIDS,* 1, 53, 1990.

Constantine, N. T., Farag, M. M., and Watts, D. M., Evaluation of a synthetic peptide-based assay and a rapid dot-blot recombinant test for the detection of antibodies to HIV-1 and HIV-2, *J. Egypt. Publ. Health Assoc.,* 115, 1, 1990.

Epstein, J., FDA Position Statement on Issues Related to Use of Combination Tests for Detection of Antibodies to HIV-1 and HIV-2, presented at the 33rd Meet. Blood Products Advisory Committee, September 26, 1991.

George, J.R., Rayfield, M. A., and Phioips, S., Efficacy of U.S. Food and Drug Administration-licensed HIV-1 screening enzyme immunoassays for detection antibodies to HIV-2, *AIDS,* 4, 321, 1990.

Gnann, J. W., McCormick, J. B., and Mitchell, S. W., Synthetic peptide immunoassay distinguishes HIV type 1 and HIV type 2 infections, *Science,* 237, 1346, 1987.

Goldstein, L. C., Epstein, J., Schochetman, G., Schable, C. A., and Zuck, T., Screening of US populations for the presence of LAV-II, *Transfusion,* 27, 542, 1987.

Mitchell, S. and Mboup, S., HIV testing, in *The Handbook for AIDS Prevention in Africa,* Family Health International, Durham, NC, 1990, 20.

Pau, C.-P., Granade, T. C., Pakekh, B., Schochetman, G., DeCock, K. M., Gayle, H., Cernescu, C., and George, J. R., Misidentification of HIV-2 proteins by Western blots, *Lancet,* 337, 616, 1991.

Parkman, P. D., Use of Genetic Systems HIV-2 EIA, Memorandum from the Food and Drug Administration, Washington, D.C., 1990.

Pollet, D. E., Saman, E. L., Peeters, D. C., Warmenblo, H. M., Heyadrickx, L. M., Wouters, C. J., Beelaert, G., van der Groen, G., and Heuverswyn, H. V., Confirmation and differentiation of antibodies to human immunodeficiency virus 1 and 2 with a strip-based assay including recombinant antigens and synthetic peptides, *Clin. Chem.,* 37, 1699, 1991.

Schumacher, R. T., Howard, J., and Ayres, L., Cross-reactivity of anti-HIV-2 positive serum in U.S. licensed screening tests for anti-HIV-1, 6th Int. Conf. on AIDS, Abstract SC627, San Francisco, June 1990, 625.

Sloand, E. M., Pitt, E., Chiarello, R. J., and Nemo, G. J., HIV testing, state of the art, *JAMA,* 266, 2861, 1991.

van der Groen, G., Van Kerckhocen, I., and Vercauteren, G., Operational characteristics of commercially available assays to determine antibodies to HIV-1 and/or HIV-2 in human sera, 6th Int. Conf. on AIDS, Abstract SC213, San Francisco, June 1990, 213.

World Health Organization, Acquired immunodeficiency syndrome (AIDS), *Wkly. Epidemiol. Rec.,* No.37, 281, 1990.

World Health Organization, Immunodeficiency syndrome (AIDS), *Wkly. Epidemiol. Rec.,* 65, 281, 1990.

Zhang, X., Constantine, N. T., Bansal, J., Callahan, J. D., and Marsiglia, V. C., Evaluation of a new generation synthetic peptide combination assay designed to detect antibodies to HIV-1, HIV-2, HTLV-I, and HTLV-II simultaneously, *J. Med. Virol.,* in press, 1992.

Chapter 6

HTLV-I AND HTLV-II

I. INTRODUCTION

At least two retroviruses other than the HIVs are known to cause human disease; they do not cause AIDS. The human T lymphotropic viruses types I and II (HTLV-I, HTLV-II) are RNA type C viruses of the subfamily Oncovirinae, that have a tropism for human T lymphocytes. Their effect is immunoproliferative rather than causing an immunodeficiency, and therefore the HTLVs are referred to as transforming viruses. The earliest human retrovirus to be discovered was HTLV-I in 1978. Subsequently, HTLV-II was identified in 1982, with HIV-1 (HTLV-III) found in 1983, finally HIV-2 (originally classified as HTLV-IV) in 1985.

It is currently estimated that >5 million individuals are infected with the HTLVs in Africa alone. The two HTLVs are related, exhibiting about 60% homology in their nucleotide sequences. Transmission of the HTLVs is by the same routes as those of the HIVs (sexual contact, through blood, and vertically from mother to infant). Transfusion is the most efficient mode of transmission, with seroconversion occurring in 35 to 60% of recipients following exposure to contaminated cellular products. In contrast to the transmission of HIV, cell-free fluids do not seem to be an efficient means of infection for the HTLVs. High rates (17%) of infection have been reported particularly in drug abusing populations in the U.S. Transmission of HTLV-I infection in transfusion recipients of infected cellular blood products is characterized by a 20 to 90 d window period, with a mean time to seroconversion of approximately 40 d. Another known mode of transmission is via breast milk. The actual pathogenic mechanisms, life cycles, and cell tropism are similar to those of the HIVs (Chapter 1). Although CD4+ T lymphocytes are the primary target for infection by the HTLVs, the CD4 receptor does not seem to be the molecule responsible for viral attachment. The highest prevalences of infection by HTLV-I are in Japan, the Caribbean, the southeastern U.S., and parts of Central America and Africa. In some areas of Japan the prevalence of infection is 5 to 10%. In nonendemic areas the prevalence is generally <0.1%. Currently, the testing of blood units for HTLV-I/II is mandated in the U.S., France, and Japan.

II. DISEASE ASSOCIATIONS

A. HTLV-I

HTLV-I has the ability to transform cells, has a long latency period before disease is manifest, and contains genes that have the ability to activate T cells. It is associated with adult lymphoma/leukemia (ATLL), HTLV associated myelopathy (HAM in Japan), and topical spastic paraparesis (TSP in the

Caribbean). It was the first human retrovirus to be associated with human malignancy. Generally speaking, no symptoms occur in early HTLV-I infection; many individuals present with no apparent disease, even after many years. It is currently estimated that the lifetime risk of developing ATLL is between 1 and 5%, while the risk of developing TSP is <1%.

ATLL is an aggressive disease marked by a proliferation of T lymphocytes, hypercalcemia, skin lesions, and opportunistic infections. Eventually, this proliferation leads to an immunocompromised state that results in fatality. TSP, also known as Jamaican myelopathy/West Indian neuropathy/HTLV-I associated myelopathy, is a progressive demyelinating neurologic disorder (similar to multiple sclerosis) that eventually leads to weakness, gait abnormalities, and neurologic abnormalities.

Following infection by HTLV-I, a long latency period ensues in which the virus appears to lie dormant, but eventually the immunoproliferation and immunocompromise result in fatality. The prognosis after diagnosis of ATLL is poor, with survival time averaging 9 weeks if hypercalcemia is present, and <2 years if hypercalcemia is not. More importantly, transfusion of HTLV-I-infected blood may be an efficient means of disease production and has been linked to the development of HAM/TSP occurring 1 month to 3 years later.

B. HTLV-II

HTLV-II infection has been described in a patient having hairy-cell leukemia; however, the agent is not conclusively linked to any specific disease. Other types of T cell lymphocyte abnormalities associated with HTLV-II infection have also been described, but viral RNA expression has never been demonstrated in malignant cells. Hairy T cell leukemia is an immunoproliferative disorder characterized by pancytopenia, and T cells that exhibit cytoplasmic projections (hairy cells).

III. ANTIGENS

A. HTLV-I

The genome of HTLV-I codes for similar gene products to those of the HIVs, i.e., gag, pol, env, and regulatory (Appendix B). The pol products are poorly described antigenically, but the *tax* gene gives rise to products that seem to have an important role antigenically. The following is a summary of the major antigenically important gene products:

gag
 p53 (precursor)
 p24/26 (core or capsid)
 p19 (matrix)
 p15
 p26, p28, p32 (gag intermediates)

env
  gp61/68 (precursor)
  gp46 (external)
  gp21 (transmembrane)
tax
  p40 (or p38) (transactivator protein)

Some similarities exist between the HIVs and the HTLVs, almost exclusively in the core region. This can, and usually does, give rise to cross-reacting antibodies (especially to p24 antigen) that will react with antigens of both viruses. Antibodies to HTLV antigens generally do not cross-react with the other antigens of HIV. Therefore, there may be some minor cross-reactivity to the p24/p26 core antigens.

B. HTLV-II

Approximately 60% of the nucleic acid sequences of HTLV-I and HTLV-II are similar. As expected, the gene products (antigenic components) of HTLV-II are similar to those of HTLV-I and are analogous to those listed for HTLV-I (above). However, the HTLV-I and -II viruses also contain distinct antigens, which may allow the viruses to be differentiated. The differences in the antigens have not been described in terms of molecular weights since different strains of the HTLVs produce antigens of a wide range of molecular weights. However, infection by each virus may produce distinct profiles by WB; this is discussed below. It may not always be possible to identify the exact infection based on routine serologic tests, and in many instances further testing by methods to evaluate the nucleic acid genome of the virus is required.

IV. SCREENING TESTS FOR HTLV-I/-II

A number of screening tests are currently available to test for antibodies to HTLV, all of which are identical in principle to those used for HIV screening (Chapter 3). The HTLV screening assays cannot differentiate infection by the two HTLV viruses since cross-reactions are omnipresent when using viral lysates; hence, the screening tests are referred to as tests for HTLV-I/-II. The extent to which HTLV-I assays detect HTLV-II antibodies and vice versa has not been quantified adequately. Since the antigens of both viruses are significantly different than the antigens of the HIVs, significant cross-reactions by antibodies do not usually occur. Therefore, tests designed to detect antibodies to the HIVs will not usually produce reactive results on serum from individuals infected with the HTLVs, and vice versa.

Similar to their use for detecting antibodies to the HIVs, ELISAs are the screening tests of choice, although alternatives such as gelatin particle agglutination exist. The ELISAs most often utilize a viral lysate-based antigen derived from the infection of T lymphocytes. The agglutination assay uses viral

TABLE 7
HTLV-I and HTLV-II Screening and Supplemental Assays[a]

Test	Manufacturer	Type of Test	Antigen
Screening Assays			
HTLV-I EIA[b]	Abbott Laboratories	I	L
Recombinant HTLV-I EIA[b]	Cambridge Biotech	I	RP
Retrotek HTLV-I ELISA[b]	Cellular Products	I	L
Detect HTLV-I/II	Coulter/IAF Biochem	I	SP
Select HTLV-I/II	Coulter/IAF Biochem	I	SP
HTLV EIA	Diagnostic Biotechnology	I	L
HTLV-I ELISA[b]	DuPont/Ortho	I	L
Serodia HTLV-I	Fujirebio	A	L
Vironostika HTLV-I EIA	Organon Tecknika	I	L
Synth EIA HTLV	UBI-Olympus	I	SP
Supplemental Assays for HTLV-I and HTLV-II			
HTLV-I/II Western blot	Cambridge Biotech	Blot	L/RP
Retro-Tek HTLV-I	Cellular Products	Blot	L
Retro-Tek HTLV-I IFA	Cellular Products	IFA	IC
HTLV-I/II Antigen Assay	Coulter/IAF Biochem	S	
HTLV Blot (2.2)	Diagnostic Biotechnology	Blot	L/RP
HTLV-I Western blot	DuPont/Ortho	Blot	L
Problot	Fujirebio	Blot	L
HTLV Western blot	Organon Tecknika	Blot	L
Sero-Fluor IFA-HTLV-I	Virion	IFA	IC
Sero-Fluor IFA-HTLV-II	Virion	IFA	IC

Key: RP = recombinant protein, SP = synthetic peptide, IC = infected cells, L = lysate, Blot = Western blot, A = agglutination, S = sandwich ELISA, IFA = indirect fluorescent assay.

[a] May not include all available tests.
[b] FDA licensed.

lysate coated particles that have no solid determinants, are biologically inactive, and incur minimal physical adsorption of unwanted serum constituents. This agglutination method is reported to be very sensitive, but specificity must be verified using control cells (Chapter 3). The agglutination test is simple to perform, can be interpreted by the naked eye, and is applicable for testing large numbers of specimens.

Similar to the tests used for detection of antibodies to HIV, HTLV-specific IgM assays have not yielded consistent results, and further evaluation is required for the determination of their utility. Recombinant antigen-based tests have been introduced and are sometimes used as screening tests. The advantages of these antigens are indicated in Chapter 3. Synthetic peptide-based HTLV tests are now commercially available and have been shown to have adequate specificity to differentiate HTLV-I and -II. Details on the operation, principles, advantages and disadvantages of the various screening assay methods are presented in Chapter 3. Table 7 lists many of the available HTLV-I/-II screening assays.

V. CONFIRMATORY TESTS AND DIFFERENTIATION OF HTLV-I AND HTLV-II

Currently, the serological methods for confirmation of antibodies to the HTLVs include WB, IFA, and RIPA (Chapter 4). Usually, one or more of these tests are required to confirm HTLV infection, although perhaps not conclusively. The manufacturers of WB kits state that the results of the WB should not be used to confirm infection or to allow individuals to return to the pool of blood donors, but only to indicate that antibodies to one or both of these viruses are present. Newer technologies, such as the synthetic peptide-based tests, may be beneficial in confirming and differentiating these infections serologically. In these tests, synthetic peptides are used that correspond to specific areas of the envelope antigens in order to differentiate the viruses; and to a conserved region of the transmembrane antigen in order to detect both viruses. Presently, viral culture and PCR are the most sensitive methods to confirm and differentiate infection by the HTLVs.

On the basis of epidemiological data, a Public Health Service Consensus Committee established criteria for HTLV-I seropositivity. A positive result requires a repeatably reactive ELISA and subsequent confirmation with a more specific serologic test. The confirmatory test must be able to identify antibody to two gene groups of the HTLV-I genome (*gag* and *env*).

Prior to 1992, WB assays could not differentiate between HTLV-I and -II infections. This was due mostly to the inadequate amount or lack of native env antigens on the blot. This may be related to the fragility of the HTLV outer membrane, which may be lost during purification. Therefore, it was common practice to perform the RIPA using type-specific viral lysates for HTLV-I and -II in order to demonstrate antibodies to the envelope components. Recently, commercial manufacturers have augmented viral lysate WBs with envelope antigens or other components in order to increase the sensitivity of blots and to assist in the differentiation of the two viruses. Some WBs now contain a unique HTLV-I envelope recombinant protein (rgp46) and/or a common HTLV-I and -II epitope purified recombinant transmembrane envelope protein (rgp21 or p21*env*r), along with the native viral antigens (lysates). However, indeterminate reactions may still occur, and profiles can be currently classified as HTLV-I/-II negative, HTLV-I positive, HTLV-II positive, HTLV-II indicative, or indeterminate. Presently, the WBs are used to monitor antibody activity to the specific viral gag proteins, p19 and p24; the env glycoproteins gp46 and/or gp61/68; and the tax protein p40 (also designated p38). Inclusion of the recombinant protein gp21 is claimed to increase the sensitivity for detecting antibodies to the envelope components, since the quantity of native lysate envelope antigens on WB may not be adequate. Sera from individuals with either HTLV-I or HTLV-II will react with this recombinant gp21, and in some cases will not react with other env antigens. The incorporation of this recombinant protein also increases the sensitivity for detection of HTLV-II. Recently, HTLV LIA tests were developed that incorporate recombinant or synthetic peptide-based antigens (Table 7) that are artificially applied to plastic strips in a WB format (Chapter 4). These tests offer ease of interpretation, and a reported increase in sensitivity and specificity.

Until recently, HTLV WBs did not contain recombinant proteins and could not be used to differentiate antibodies to the two viruses. Criteria for positivity required reactivity to p19 or p24; and to an envelope antigen (gp46 or gp68). A positive result indicated that the individual was infected by one or the other of the viruses. If an indeterminate result was produced, e.g., reaction to only p19 or p24, then a RIPA had to be performed in order to demonstrate reactivity to the envelope antigens. This was unfortunate, since the RIPA was only performed in selected laboratories that had the appropriate facilities.

At the time of writing, the HTLV WBs that may be helpful in differentiating HTLV-I and -II antibodies are those from Diagnostic Biotechnology and from Cambridge Biotech. However, differentiation has not always correlated with known positives, and therefore results must still be interpreted with caution. For both of these blots, the criteria for a sample to be classified as negative and as indeterminate are common: HTLV-I/-II negative — no reactivity to viral specific bands or to rgp21 or rgp46 (if present); indeterminate — any reactions present, but that do not meet the requirements for the classification of positive.

The criteria for a positive WB by the two companies are similar to one another and are noted in Table 8.

Interpretation of bands on HTLV blots is difficult. They must be compared to the positive controls in all cases, but are still more difficult than interpreting HIV blots. Figure 41 is a representation of HTLV-I and -II positive reactions by WB from Diagnostic Biotechnology and Cambridge Biotech.

Similar to the results generated by HIV assays, a negative result does not exclude the possibility of infection by HTLV, and an indeterminate result is inconclusive and should not be interpreted as evidence of infection. Dually infected specimens cannot be conclusively determined using the above criteria, but may yield a result classified as HTLV-I positive.

Nonspecific reactions in noninfected individuals can occur, particularly to p19 and rgp21. Rarely, a nonspecific reaction to p24 and rgp21 may occur, yielding a classification of HTLV-II indicative (Diagnostic Biotechnology) or HTLV positive (Cambridge Biotech), although the individual is not infected. Reactivity to rgp46 is very specific in HTLV-I-infected individuals. Therefore, when reactivity to this recombinant glycoprotein occurs, along with reactivity to a gag and rgp21, HTLV-I infection is confirmed (Diagnostic Biotechnology). However, some individuals may be infected with HTLV-I but have no reactivity to rgp46; i.e., the rgp46 reaction is not 100% sensitive. Several reports indicate that sera from HTLV-I-infected individuals react more strongly

TABLE 8
Criteria for HTLV-I- and HTLV-II-Positive Western Blots from Two Commercial Companies

	Diagnostic Biotechnology	Cambridge Biotech
HTLV-I positive	gag (p19 or p24) env (gp46 or rgp46) and rgp 21	
HTLV-I or -II positive		gag (p19 or p24) and env (gp46 or p21envr)
HTLV-II indicative	At least p24 and rgp21, but not rgp46	

Notes: (1) rgp21 and p21envr are different terminologies for the recombinant p21 protein; (2) it is suggested by both companies that reactivity to p19 is greater than to p24 in HTLV-I infection, while reactivity to p24 is greater than to p19 in HTLV-II infection; (3) both manufacturers recommend additional testing by other technologies; (4) the above criteria were obtained from the package inserts from the manufacturers; (5) it is likely that these criteria will change.

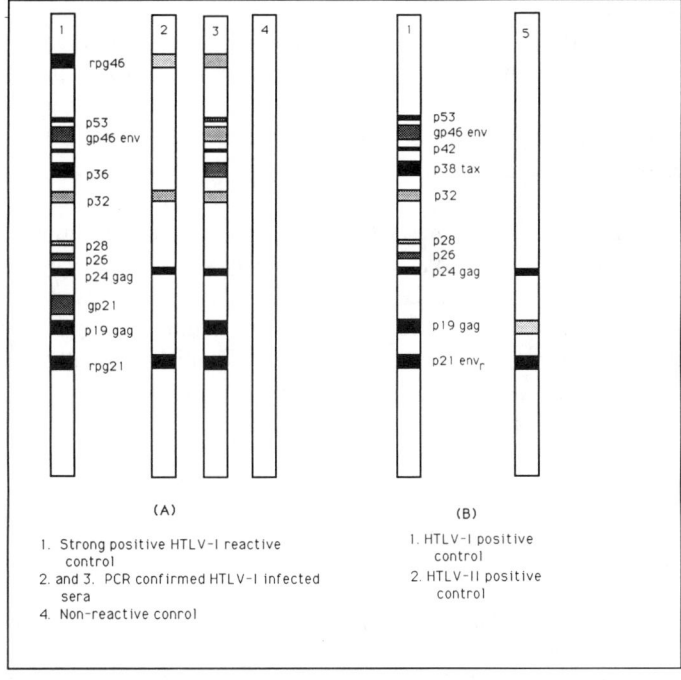

FIGURE 41. Positive reactions by HTLV WBs from (A) Diagnostic Biotechnology and (B) Cambridge Biotech.

with p19 than with p24; and sera from individuals infected with HTLV-II react more strongly with p24 than with p19. In fact, one report indicates that this observation can be used to predict the specific virus with 100% efficiency (providing that reactivity to rgp21 also occurs). Recently, much interest has arisen in reactivity to the *tax* gene product, p40 (or p40x or p38). In most cases of HTLV-I infection, antibodies to p40 can be demonstrated.

Currently, the FDA criteria for a positive HTLV-I WB require the presence of bands from two of the three gene products (gag, env, and tax). The absence of all bands constitutes a negative result, while reactions to any bands not meeting the criteria for positivity are classified as indeterminate. The criteria established by WHO for a sample to be confirmed by WB include reactivity to one gag protein (core or matrix) and one envelope antigen (precursor or envelope). The Public Health Service Working Group has established criteria to define positivity that include reactivity to the p24 (or rarely p19) gag protein and at least one env protein (gp46 or gp61/68).

Samples yielding indeterminate results by WB should be tested by alternative supplemental tests such as IFA and/or RIPA. A recent report has indicated that a significant number of individuals whose sera are classified as indeterminate by viral lysate WBs are truly infected. In addition, some individuals who were confirmed positive for HTLV-II DNA by PCR had WB profiles that did not include reactions to p24 or rgp21, suggesting that false-negative results may occur by WBs.

The IFA and RIPA techniques are similar to those performed for HIV, with the exception of the infecting virus (Chapter 4). In the IFA technique, which uses infected cells such as MT-2, a positive test is recorded when a diffuse cytoplasmic or a capping type of fluorescence appears. A significant number of indeterminants occur, usually by the production of weak reactions. Some cross-reactions from HIV infected individuals also occur.

RIPA is considered by some authorities to be the most sensitive assay to detect antibody to the envelope antigens, namely gp61/68 (actual molecular weight depends on the infected cell line) and gp46. As previously stated, RIPA is a research method and is not generally used in routine testing laboratories. Recently, a test was developed to detect HTLV-I antigen (usually p19). The utility of this assay for testing serum has yet to be determined.

REFERENCES

AIDS 89 Summary. A Practical Synopsis of the Fifth International Conference, Philadelphia Sciences Group, Philadelphia, 1989, 1.

Alexander, S., HTLV-I Confirmatory Testing, Internal Publication, American Association of Blood Banks, 4, 4, 1991.

Allain, J. P., Gallo, R. C., and Montagnier, L., Human Retroviruses and Diseases They Cause, Abbott Laboratories/Excerpta Medica, Amsterdam, 1988, 1.

Committee on Human Retrovirus Testing, Fifth Consensus Conference on Testing for Human Retroviruses: Report and Recommendations. Kansas City, MO, 1990, 5.

Diagnostic Biotechnology, HTLV BLOT 2.2 package insert, Singapore, April 1991.

Hjelle, B., Mills, R., Cyrus, S., and Swenson, S. G., Infection is present in a high fraction of HTLV western blot (WB)-indeterminate donors, *Transfusion,* 31, 8S, 1991.

Kiyokawa, T., Yamaguchi, K., Nishimura, Y., Fukuoka, N., Kline, R. L., Brothers, T., Halsey, N., Boulos, R., and Watanabe, T., Western Blot criteria for HTLV-I, *Lancet,* 338, 312, 1991.

Lairmore, M. D. and Quinn, T. C., Evaluation of enzyme immunoassays for antibody to human T-lymphotropic viruses type I/II, *Lancet,* 337, 30, 1991.

Lentino, J. R., Pachucki, C. T., Schaaff, D. M., Schaefer, M. R., Holzer, T. J., Heynen, C., Dawson, G., and Dorus, W., Seroprevalence of HTLV-I/II and HIV-1 infection among male intravenous drug abusers in Chicago, *J. AIDS,* 4, 901, 1991.

Lillehoj, E. P. and Alexander, S. S., Antibodies to human T-cell leukemia virus type I transactivator protein detected by whole virus enzyme immunoassay, *Transfusion,* 31, 8S, 1991.

Lillehoj, E. P., Tai, C. C., Nguyen, A., and Alexander, S. S., Characterization of env and tax encoded polypeptides of human T-cell leukemia virus type I, *Clin. Biotechnol.,* 1, 27, 1989.

Lillo, F., Varnier, O. E., Sabbatani, S., Ferro, A., and Mendez, P., Detection of HTLV-I and not HTLV-II infection in Guinea Bissau (West Africa), *J. AIDS,* 4, 541, 1991.

Manns, A. and Blattner, W. A., The epidemiology of the human T-cell lymphotrophic virus type I and type II: etiologic role in human disease, *Transfusion,* 31, 1, 1991.

Centers for Disease Control, Licensure of screening tests for antibody to human T lymphotropic virus type 1, *MMWR,* 37, 736 and 745, 1988.

Satow, Y., Ishikawa, K., Mizuno, M., Hashidol, M., Honda, H., Hiroshi, K., and Hayawi, M., Detection of HTLV-I antigen in peripheral and cord blood lymphocytes from carrier mothers, *Lancet,* 338, 915, 1991.

Shih, J., O'Donnell, D., Delancey, S., Mann, T., and Lee, H., Differentiation of HTLV-II using synthetic peptides from immunodominant regions of HTLV env proteins, *Transfusion,* 31, 8S, 1991.

Viscidi, R. P., Hill, P. M., Li, S., Cerny, E. H., Vlahov, D., Farzadegan, H., Halsey, N., Kelen, G. D., Quinn, T. C., Diagnosis and differentiation of HTLV-I and HTLV-II infection by enzyme immunoassay using synthetic peptides, *J. AIDS,* 4, 1190, 1991.

Wiktor, S. Z., Alexander, S. S., Shaw, G. M., Weiss, S. H., Murphy, E. L., Wilks, R. J., Shortly, V. J., Hancard, B., and Blattner, W. A., Distinguishing between HTLV-I and HTLV-II by Western blot, *Lancet,* 335, 1533, 1990.

Wiktor, S. Z., Pate, E. J., Weiss, S. H., Gohd, R. S., Correa, P., Fontham, E. T., Hanchard, B., Biggar, R. J., and Blattner, W. A., Sensitivity of HTLV-I antibody assays for HTLV-II, *Lancet,* 338, 512, 1991.

Chapter 7

INDICATORS OF THE VALUE OF DIAGNOSTIC TESTS

I. INTRODUCTION

It is appropriate to once again question which test is the best. Obviously, no best test exists, otherwise, all laboratories would use that test exclusively. Of course, it depends on the need in the specific testing situation, and the characteristics of the individual tests. Certain tests perform better in some geographic locations. For example, sera from one location may contain antibodies to a variety of parasitic organisms found in that area, and therefore may produce a significant number of false-positive results using a particular test format. Differences in test performance have been reported depending on the origin of the sera. The value of some tests may depend on the expertise of the laboratory workers. If a laboratory worker has not had the proper training for conducting the IFA, for example, the results may not compare favorably with other tests, and therefore the IFA will be considered an inferior test. Certain tests may not perform optimally if the conditions in the laboratory are not optimal (e.g., ambient temperature varies from that which is required). General guidelines for conducting evaluation studies and controlling for these differences are described in Chapter 10.

Parameters are available that can be used to determine the relative usefulness or efficiency of a particular test, assuming that each test has been evaluated by properly trained individuals, performed exactly as required by the manufacturer, evaluated on the same population of sera, and performed under the same testing conditions. The test parameters that can assist in determining the usefulness of a test are sensitivity, specificity, efficiency, delta values, and predictive values.

II. SENSITIVITY

"Sensitivity" of a diagnostic assay can have two different meanings. It can indicate the ability of a test to detect very small amounts of the analyte (e.g., antibody). For example, if one test can detect antibody while another cannot, the former is more sensitive. Similarly, if one test produces a titer of 256 while a second test only yields a titer of 64, the first test detects more antibody and is more sensitive. Many laboratories will dilute positive sera and try to determine which test is more sensitive by observing the ability of the test to detect smaller quantities of antibody. This would be indicative of the ability for the test to detect seroconversion at an earlier time. The second meaning of sensitivity is related to the first, but is the ability of a test to detect truly infected individuals. By detecting all infected individuals the test will not produce false-negative results. This differs from the first definition in that a test may be able to detect very small quantities of certain antibodies, but may still miss some infected individuals due to the absence of the proper antigens incorporated in the test.

The sensitivity of an assay can be calculated using the following formula in which true positives and true negatives relate to the infected or noninfected individuals that the test correctly identifies:

$$\text{Sensitivity} = \frac{\text{True Positives}}{\text{True Positives} + \text{False Negatives}} \times 100$$

Example
Population:
100 sera are tested
5 sera are from infected individuals
95 sera are from noninfected individuals

Test results:
In comparison to the reference test, the test reveals only 4 positives among the sera from the 5 infected individuals (the test produces 1 false-negative).

The sensitivity of the test is thus calculated to be 80%:

$$\frac{4}{4+1} = 0.80 \times 100 = 80\%$$

III. SPECIFICITY

The specificity of an assay is the ability of the test to identify all noninfected individuals correctly (i.e., produces no false-positive results). Most screening assays for HIV are not 100% specific, and therefore results must be verified using a more specific supplemental test. The specificity of an assay can be calculated from the following formula:

$$\text{Specificity} = \frac{\text{True Negatives}}{\text{True Negatives} + \text{False Positives}} \times 100$$

Example
Population:
100 sera are tested
5 sera are from infected individuals
95 sera are from noninfected individuals

Test results:
   In comparison with the reference test, the test reveals 6 positives (all 5 of the infected individuals and 1 false-positive from the noninfected group). Therefore, the test correctly identified 94 of the noninfected individuals (produced 94 true negatives), and 1 false-positive.

The specificity of the test is thus calculated to be 98.9%:

$$\frac{94}{94 + 1} = 0.989 \times 100 = 98.9\%$$

IV. TEST EFFICIENCY

Test efficiency refers to the overall ability of a test to correctly identify all positives and negatives (the absence of false-positives and false-negatives). It is a combination of the sensitivity and the specificity of an assay that determines the total effectiveness of the assay. It is determined as follows:

$$\frac{\text{True Positive} + \text{True Negative}}{\text{True Positive} + \text{False Positive} + \text{True Negative} + \text{False Negative}} \times 100$$

Example
Population:
   100 sera tested
   5 sera from infected individuals
   95 sera from noninfected individuals
Test results:
   5 positives (4 from the infected group and 1 false positive from the noninfected group).
   95 negatives (94 from the noninfected group and 1 false-negative from the infected group).

The test efficiency is thus calculated to be 98%:

$$\frac{4 + 94}{4 + 1 + 94 + 1} = 0.98 \times 100 = 98\%$$

V. DELTA VALUES

Although most tests are evaluated in terms of sensitivity and specificity, quantitative differences can be better determined by noting actual values generated by the tests. Sensitivity and specificity relate to purely qualitative, retrospective evaluation since they are based on whether a sample was positive or negative and the patient determined to be truly infected or not infected. When evaluating tests such as ELISAs that generate O.D. values, it can be valuable to know if a false-positive result was close to the cutoff in one test while being strongly positive by another test. In the case that a false-positive result lies in an area in which most truly positive results lie, then the test may be fundamentally flawed. Therefore, it may be helpful to employ a measure that summarizes how effective a test is in separating positive and negative populations. An approach to determine statistical estimates of sensitivity involves the use of calculated delta values.

Delta values are defined as the distance of the mean O.D. ratio of the sample population from the cutoff measured in standard deviation units, and are calculated by dividing the mean O.D./cutoff ratio (log 10) of the samples by the standard deviation. This is performed on the positive samples for the positive delta value, and on the negative samples for the negative delta value.

The following is an example of calculations for delta values: 3 positive and 3 negative samples are tested by an indirect ELISA. From the O.D./C.O. ratios the positive and negative delta values are calculated as follows:

	Positive samples	Negative samples
O.D./C.O.	3.10	0.04
	3.40	0.01
	4.20	0.02
Mean O.D./C.O.	3.56	0.02
Standard deviation	0.46	0.01
Log_{10} of the mean O.D./C.O.	0.55	−1.69
Delta values	Positive delta value	Negative delta value
	+1.20	−169

When two tests are being compared, the test that has a higher positive delta value would better characterize positive samples; the test with a higher negative delta value would better classify negative samples. For clarity, a test exhibiting a negative delta value of −160 represents better resolution than a test with a negative delta value of −140. When using delta values for evaluating tests, the values should be determined using a large number of samples (at least 50) to ensure that the test population is well represented.

VI. PREDICTIVE VALUES

Predictive values differ from the above parameters in that they describe the value of tests, taking into account the actual prevalence of infection in the population being tested. Therefore, the value of a test may not depend on its sensitivity and specificity as much as it depends on the population being tested. The positive predictive value (PPV) is the frequency of infected individuals among all persons with positive results. The negative predictive value (NPV) is the frequency of noninfected individuals among all persons with negative results. An example illustrates the point. The predictive values are calculated as follows:

$$\text{Positive Predictive Value} = \frac{\text{True Positives}}{\text{True Positives} + \text{False Positives}} \times 100$$

$$\text{Negative Predictive Value} = \frac{\text{True Negatives}}{\text{True Negatives} + \text{False Negatives}} \times 100$$

Population #1, where the prevalence of infection is high (5%)

Population:
  1000 sera tested
  50 sera from infected individuals
  950 sera from noninfected individuals

Test results:
  50 positives (45 from the infected group and 5 false-positives from the noninfected group)
  950 negatives (945 from the noninfected group and 5 false-negatives from the infected group)

Therefore, the predictive values are

$$\text{PPV} = \frac{45}{45 + 5} = 90\%$$

$$\text{NPV} = \frac{945}{945 + 5} = 99.5\%$$

With the same test results (same number of false-positives and false-negatives) in a different population, the following is noted.

Population #2, where the prevalence of infection is low (0.7%)

Population:
    1000 sera tested
    7 sera from infected individuals
    993 sera from noninfected individuals

Test results:
    7 positive results (2 from the infected group and 5 false-positives from the noninfected group)
    993 negatives (988 from the noninfected group and 5 false-negatives from the infected group)

Therefore, the predictive values are

$$PPV = \frac{2}{2 + 5} = 28.6\%$$

$$NPV = \frac{988}{988 + 5} = 99.5\%$$

As is dramatically shown, the same test that yields the same number of false-positives and -negatives produces a different positive predictive value when testing two different populations.

The specificity of the test in population #2 is an excellent 99.5% (988/988 + 5); however, the chance of a positive result being from a truly infected individual is only 28.6% (2 true positives detected by the test and 5 false-positives). This indicates that a positive result by the test will come only from an infected individual in one of four cases (a guess could yield a better chance!). Therefore, the specificity of the test alone does not indicate the usefulness of the test in this low prevalence population; the predictive value must be determined.

REFERENCES

Crofts, N., Maskill, W., and Gust, I. D., Evaluation of enzyme-linked immunosorbent assays: a method of data analysis, *J. Virol. Methods*, 22, 51, 1988.

Griner, P. F., Mayewski, R. J., Mushlin, A. I., and Greenland, P., Selection and interpretation of diagnostic tests and procedures, *Ann. Intern. Med.*, 94, 557, 1981.

Maskill, W. J., Crofts, N., Waldman, E., Healey, D. S., Howard, T. S., Silvester, C., and Gust, I. D., An evaluation of competitive and second generation ELISA screening tests for antibody to HIV, *J. Virol. Methods*, 22, 61, 1988.

Report of the WHO Meeting on Criteria for the Evaluation and Standardization of Diagnostic Tests for the Detection of HIV Antibody, Stockholm, WHO/GPA/BMR/88.1, World Health Organization, Geneva, 1987.

Chapter 8

QUALITY CONTROL AND QUALITY ASSURANCE

I. NECESSITY AND IMPORTANCE

The validity of diagnostic test results produced in each laboratory is entirely dependent on the quality of the measures employed before, during, and after each assay. Consistency in the production of good results requires an overall program that includes quality assurance (QA), quality control (QC), and quality assessment. Most importantly, inaccurate results due to technical or transcriptional errors are inexcusable, and can be prevented by a good QA program.

Measures to control the quality of the results in an HIV diagnostic laboratory are extremely important, because the consequences of either a false-positive or a false-negative result are great:

- A person falsely labeled as negative for antibodies to HIV may unknowingly continue to infect other individuals. Therefore, each infected person who tests negative has the potential for creating his or her own epidemic.
- A person who is falsely labeled as positive may be subjected to all types of discrimination, may lose employment, may be denied health insurance benefits, may be cast out by family, and in some cases may even commit suicide.
- Legal considerations also accompany erroneous results. Currently, several lawsuits pending throughout the U.S. involve the transfusion of infected blood units. Most of these cases represent transfusions occurring before federally mandated HIV testing regulations were instituted in 1985. However, the outcome of these cases could very well affect laboratory policy standards by establishing new precedents for liability if negligence by a testing facility can be demonstrated.

Thus, the effectiveness, reputation, and possibly even the accreditation of the laboratory are dependent upon the quality of work generated.

Many variables can (and will) affect the quality of results:

- The educational background, certification, and training level of the laboratory personnel
- The condition of the specimens
- The controls used with the test runs
- The interpretation of the results
- The transcription of results
- The reporting of results

II. CONFIDENTIALITY

Specific HIV laboratory results and the names of people tested (regardless of results) should never be a topic of loose discussion. The privacy and rights of an individual can be severely compromised by information from overheard conversation. HIV testing results should not be available for general viewing, and must be kept in a secure location to prevent access by unauthorized individuals. Laboratory reports should be submitted in a sealed envelope marked "confidential", then hand delivered to the submitting clinician to preserve confidentiality. The authors do not recommend that results be communicated via telephone or by computer, so that confidentiality is not compromised.

III. DEFINITIONS

A. Quality Control
Quality control (QC) refers to those measures that must be included during each assay to verify that the test is working properly.

B. Quality Assurance
Quality assurance (QA) is defined as the overall program that ensures that the final results reported by the laboratory are correct (as accurate as possible).

C. Quality Assessment
Quality assessment (also known as proficiency testing) is a means to determine the quality of the results generated by the laboratory. It is usually an external evaluation of a laboratory's performance, involving the incorporation of proficiency panels as the means of evaluation. Internal quality assessment programs can also be instituted. Quality assessment is a challenge to the effectiveness of the QA and QC programs.

IV. QUALITY ASSURANCE: FUNDAMENTALS FOR OVERALL QUALITY OF RESULTS

QA is an ongoing process that requires daily attention by all laboratory and related staff. The responsibility and obligation for maintaining the highest testing standards require a team effort and must include every parameter affecting test results and final reports. Some fundamental issues in QA related to specimens include:

- All specimens to be tested must be inspected upon receipt, and before testing, to ensure that they are suitable. Lipemic, hemolyzed, or contaminated samples should not be used since these sample conditions may

interfere with assay performance, thus yielding questionable results. If such a specimen is received, notify the submitter and request a new specimen.
- Serum or plasma are the specimens of choice for antibody or antigen assays. Most commercial kits, unless otherwise indicated, are not suitable for use with CSF or body fluid samples.
- Before accepting a specimen in the laboratory, ensure that the tube of blood received is properly labeled. Information on the specimen label should include the full patient name at the minimum, but preferably also the collection date, submitting physician, and patient identification number, if applicable. Some laboratories save the original tube for several days or longer for retesting in the case of a positive HIV result.
- If a serum sample is frozen before testing, the sample must be well mixed after thawing and before testing.
- All test results, controls, and records must be checked and rechecked to ensure that the proper specimen was resulted before reporting.

QA also includes such factors as:

- Reporting results in a timely manner
- Being sure that the results are reported to the appropriate individual
- Making sure the laboratory is functioning in the most efficient way
- Including a continuing education program for laboratory workers
- Evaluation of laboratory personnel to identify areas for improvement
- Using the most reliable tests
- Reviewing transcriptional measures
- Verifying final reports

The following sections indicate components that must be continually monitored, and represent fundamental aspects of a good QA program.

A. Record Keeping

An efficient laboratory will be able to monitor the records of specimens from the time the samples arrive until the time that results are released. Log books are an essential step in the recording of laboratory specimens. It is imperative that log books containing names be kept confidential. The log book should include:

1. The name and/or specimen identification
2. The date of sample collection
3. The date received in the laboratory
4. The name of the requesting physician
5. The specific tests requested

Any specimen that is determined to be inadequate for testing or that does not contain the essential information (including labeling), should not be tested and a note should be entered in the log book. The submitter should be notified immediately.

A worksheet (platemap) that indicates the location of each test sample on the plate must accompany each test run performed in the laboratory. The worksheet should be filled out before the first sample is added to the plate, which then serves as a guide when placing samples in the run. The worksheet should help reduce confusion and errors during testing, although it cannot eliminate errors such as adding specimens to the wrong well or forgetting to add a sample. Sample addition to the plate always requires undivided attention. Each worksheet should contain the kit lot number, expiration date, date performed, and the technician's initials. Completed worksheets with results can be filed for a permanent record. If available, records can be stored in a laboratory computer system (a backup disk or hard copy of results should always be kept).

QC records are important in validating laboratory results. General requirements for certification of laboratories are not addressed in this book. However, the "Accreditation Manual for Hospitals (AMH), Volumes I and II", can be purchased from the Joint Committee for Accreditation of Health Care Organizations (JCAHO) at 1 Renaissance Blvd., Oakbrook Terrace, IL 60181 (708) 916-5600.

All licensing agencies in the U.S. require that records be retained for several years. During inspections, these agencies may ask to see all records. Daily log sheets for temperatures of water baths, incubators, refrigerators, and freezers document that the operating conditions used to store and run the kits and samples were maintained appropriately.

A standard operating procedure (SOP) manual should be kept in the laboratory at all times. The SOP manual contains detailed explanations of all procedures as they are performed in the laboratory. This can act as a quick reference for any procedure, and as a teaching source for new employees and students. The SOP should be reviewed and updated frequently. Periodically, the lab manager should monitor technicians to ensure that they are following the SOP. Package inserts for kits can be placed in a book, but cannot substitute for the SOP. The SOP should be so detailed that a technologist unfamiliar with the test can follow the instructions and perform the assay properly.

B. Monitoring Laboratory Staff

Laboratory managers may wish to periodically monitor the performance of their laboratory workers. Managers may establish a system wherein samples with known results (or samples already tested) are resubmitted discreetly along with the routine workload. To be effective, the laboratory workers should not be made aware of when these samples are submitted; thus, the samples are not given special treatment. This is done so that the laboratory manager, who is ultimately responsible for all results reported from the laboratory, can monitor

Quality Control and Quality Assurance

the actual performance of the laboratory staff. This monitoring should not be done to intimidate or trick the workers, but rather to help them feel confident about their results. They should be reminded that blind samples will be included periodically in their workload.

C. Vigilance in the Laboratory

Vigilance refers to watchfulness. Scrutiny of every aspect of the laboratory and every step during the testing of specimens will help ensure that the final results will be reliable. This begins when the specimens and requisition slips are received in the laboratory, and ends when the final report is released to the clinician. Particularly important is the ability to recognize a mistake or problem and report it to other workers and the supervisor. Of course, corrective action must then be taken.

The testing and the accurate reporting of results is of primary concern when testing for HIV. To fulfill this obligation, a QA program is of utmost importance. QA measures will help ensure reliability of results; however, reliability also requires vigilance on the part of every laboratory worker. The laboratory must function as a team, with each worker taking the responsibility for monitoring their own and their co-workers' performance, checking and rechecking to ensure that all results are reported correctly. QC discrepancies must be brought to the attention of the supervisor. Vigilance consists of always: (1) watching that the identification on the specimen matches that of the requisition slip; (2) noting the condition of specimens as they are received (i.e., are they adequate for testing; are they blood, urine, or CSF); (3) reviewing QC charts each day, noting small changes that may indicate the development of a problem; (4) observing co-workers, and identifying potential problems or inaccuracies, and discussing these matters in a diplomatic manner; (5) taking care that the laboratory remains a safe place to work; (6) rechecking paperwork and worksheets before reporting a result.

D. Verification of True Positives and True Negatives

A positive HIV result is a serious concern. Each laboratory must be absolutely certain that each positive result is correct. Once a sample is found to be "reactive" by a screening test, an aliquot from the initial specimen tube should be retested. This will help to eliminate handling or labeling mistakes incurred by any subsequent aliquoting. Whenever possible, a second specimen should be collected from the individual and retested to eliminate any possible handling, labeling, or clerical errors. Also, to be certain that the sample is truly reactive, the specimen can be retested using an assay based on an alternative principle, or a supplemental test.

A current trend in HIV testing is to classify a sample as "reactive" by a screening test and then "positive" only after verifying the result by testing with a confirmatory assay. In most cases a repeatedly reactive result by a screening assay and a positive result by a confirmatory assay are sufficient to verify positivity; however, the sample may be tested using a second confirmatory assay, if desired. For example, assuming that all controls were valid in both the screening and confirmatory assays, if there is any reason whatsoever that causes doubt that the confirmatory results are correct, a second confirmatory assay can be used to verify positivity. If a laboratory uses an IFA to confirm, then a WB can subsequently be used to verify positivity.

Verification of negative results is also important, especially in screening blood for transfusion. A representative sample (10%) of initially nonreactive specimens may be retested to verify that false-negative results did not occur.

E. Parallel Testing of Resubmitted Specimens

Parallel testing of resubmitted specimens is important in those patients whose specimens have yielded indeterminate results. The original specimen is held until the resubmitted specimen is received some time later, and then both specimens are tested in parallel during the same run, observing for changes in reactivity. The results of both tests can easily be compared.

F. Reviewing Transcriptional Measures

Reporting of erroneous results due to transcriptional errors is inexcusable. Transcriptional or clerical errors include mistakes made during the transfer of information from the test readout to the worksheet, and from the worksheet to the computer or report form. These types of errors probably account for the majority of errors that occur in the laboratory. Therefore, it is imperative that a system be developed to recheck results at each of these steps. Two possible mechanisms to address transcriptional errors include: (1) have a second technologist read the results from the instrument/worksheet/computer to another technologist who will check the final result and (2) have a supervisor check all results at the end of the day, before releasing the results from the laboratory. Ideally, both of these measures should be instituted, if time permits. Again, most mistakes that occur are due to transcriptional errors.

G. Reporting of Results

The nature of HIV infection and the impact on the individual and society are important issues, therefore the handling of results must be controlled so that the confidentiality of all persons tested is protected. A policy decision on the handling of HIV test results must be established and uniformly enforced in any laboratory in which HIV testing is performed. Ideally, HIV test results are reported to the submitting physician, who in turn can appropriately inform and counsel the tested individual.

Assuming that there is flawless performance in the testing process and a correct interpretation has been made regarding the HIV result, an incorrect or misleading report can still cause problems in the HIV diagnostic system. Examples of reporting inaccuracies include: the use of outdated nomenclature for HIV-1 such as HTLV-III, which can be easily confused with HTLV-I or

HTLV-II; the use of the term LAV, which is a former designation of HIV-1 in certain locations; reporting of reactive HIV ELISA screening results without confirmation on a report form labeled "Final Report" (i.e., reactive ELISA results are reported as final); the use of obsolete WB interpretive criteria; or, reporting "HIV ANTIBODY POSITIVE" without designating whether the result is from an ELISA, IFA, or WB. Any of these situations can confuse the clinician and may result in inappropriate counseling or treatment, representing a conspicuous failure of the HIV diagnostic system.

When an indeterminate result is reported, care should be taken to ensure that the clinician understands the significance of such a result and the importance of a follow-up specimen. Results should be clearly indicated on the report form and should provide comments as necessary. Figure 42 shows an example of a report sheet that could be used to communicate results.

H. Interaction with Physicians

The relationship between laboratory personnel and physicians should be one of mutual trust and respect. Communication is important in establishing rapport, since many times there will be questions about interpreting laboratory results. In some instances, the laboratory worker may know more about the significance of the laboratory findings than the physician. The laboratory must earn trust and respect by being understanding and courteous while explaining the assay principles, interpretation, and significance of the results. This can be a time of mutual education for both parties, as the clinical and laboratory results are assessed for their consistency.

I. Storage of Specimens for Follow-Up Testing

Organization in the laboratory is essential. When a blood specimen is submitted for HIV testing, all information is entered into the laboratory log book, and the specimen is centrifuged. The serum or plasma is then separated into a vial suitable for testing and storage. Ideally, this vial should be able to hold at least 1 ml of serum, a volume more than adequate for screening and confirmation. Prior to testing, samples may be stored up to 1 week at 4°C or until all testing is completed. Once the sample has been tested, it may be stored at –20°C for years. A serum bank can be established in which frozen samples are neatly stored in order by accession number. A log book containing all pertinent information should be capable of identifying all stored samples. Samples should not be repeatedly thawed and refrozen, because this process can denature proteins and lead to inaccurate results. If repeated testing is anticipated, the contents of the original vial may be subdivided into several vials and identically labeled in order to preserve sample integrity.

For specimen storage, 1-ml plastic screw cap O-ring vials are recommended. Glass vials may eventually crack with long-term frozen storage and certainly are subject to breakage if dropped. The capacity of the storage container should match as closely as possible the volume of the serum; other-

```
                    HIV Testing Result
                       Confidential

Patient Name
Patient history
Patient location
Date collected


Physicans Name  ........................    Reviewed by  ...........................


Elisa Result

      * Non-Reactive

      * Repeatedly Reactive

      * Repeatedly Gray Zone ( negative value but within 20% of cutoff; this may
      indicate that antibodies are present at low level, such that occurs during
      seroconversion)

Western Blot Result :

      * Negative ( no bands)
      * Positive
          Bands Present:

            p17  p24  p31  gp41  p51  p55  p66  gp120  gp160   All

      * Indeterminate
          Bands present:

            p17  P24  P31  GP41  P51  P55  P66  GP120  GP160

        Patients with indeterminate results should be followed after one month for
      evidence of seroconversion. Please submit follow-up specimens and indicate
      such, so that the original and the new specimen can be tested in parallel.


      COMMENT :  ..........................................................

      NOTES :

            * It is assumed that the physican has obtained informed
      consent from the patient before requesting HIV tests. Please be
      sure that all requests for HIV testing contain a physician's signature.
```

FIGURE 42. An example of an HIV report form.

wise, evaporation and concentration of the serum will occur during long-term frozen storage. For example, do not store 50 µl of serum in a 5-ml tube.

When labeling storage vials, consider using typed or computer-generated labels with one sticky surface, rather than writing directly on the vials. Another alternative is to use waterproof, fine-point marking pens. The labeling must adhere firmly at all times. As samples are removed from the refrigerator or freezer and brought to room temperature, condensation may develop, handwritten numbers are easily smudged beyond recognition, and some labels may fall off. If printed labels are unavailable, a piece of transparent tape over the number prevents smudging, and helps assure adherence.

Once a sample has been tested, it should be stored at $-20°C$ in the appropriate vial. An organized serum bank can serve many useful purposes:

- Providing a ready source of sera to assemble panels of known negative, positive, and indeterminate samples for evaluation of new test methods or kits as they become available.
- In some locations, retrospective epidemiological surveys for agents other than HIV may be performed. Studies of this type are facilitated by access to accurate records that match properly cataloged sera. A source of characterized sera is then available for testing as new infectious agents are identified.
- Following up and monitoring individuals for seroprogression by testing previously collected sera along with current sera can be accomplished.

Incidents may occasionally occur in which the results of a follow-up sample do not match the initial reactive results. For this reason, some laboratories find it valuable to retain their ELISA reactive specimens in the original labeled tubes (clot tubes) that were submitted for a period of 4 to 6 months at $4°C$. Long-term storage at this temperature usually does not alter truly reactive results.

J. Laboratory Efficiency

It is possible to have a redundant system in an efficient laboratory. Redundant measures require extra time, but they are necessary to ensure maximum reliability. For the quickest turnaround time for HIV reports, every step of the process must be well defined and work assignments organized accordingly. Arrange priorities so that blood tubes can be centrifuged, aliquoted, and labeled as soon as is possible after they arrive in the laboratory. Schedule HIV runs on as many days of the week as is financially feasible.* The supervisor should

---

* A run is any number of plates (of the same lot) assayed simultaneously by the same technologist/technician. Many laboratories batch specimens to save reagent and costs. No maximum number of specimens has been set that can be included on a run; however, when performing a larger run, do not attempt to add more specimens than can be adequately managed within the stated incubation times. The approach in handling plates with a 96-well format is slightly different than when performing bead-type assays. The 96-well plate has antibody or antigen bound to the well and thus extreme variations in time between the first and last sample addition should be avoided; therefore, the authors recommend a two-plate maximum for each separate run. With the bead-type format, the reaction begins when the bead is added; also, devices are available that enable addition of all beads to a single plate simultaneously. Therefore, larger runs can be attempted with bead-type assays as long as the wells of each plate are covered after sample addition and before proceeding to the next plate (to prevent evaporation and concentration of sample). Each run must have a full set of controls, one full set on the first plate and at least one of each control for each subsequent plate.

review HIV results at the end of each run or day so that appropriate reports can be sent out, repeat samples can be requested, and retesting can be performed promptly. This will assure minimum turnaround time for submitted samples. Always allow time to review worksheets, computer entries, and reports for transcriptional and clerical errors.

Open communication with the testing staff is also important to encourage suggestions that may further modify and streamline the system. Laboratory efficiency requires motivation and cooperation from all staff members.

As HIV testing technology expands, the laboratory cannot afford to become entrenched in an unchanging routine from year to year. Other items that must be periodically evaluated to increase efficiency in the laboratory include:

- Are the best available tests being used?
- Are the most economical tests being used?
- Are newer tests being continually evaluated?
- Is the computer system efficient and effective?
- Is the laboratory budget properly divided and utilized?

K. Total Quality Management
1. Introduction

Total quality management (TQM) refers to a comprehensive organizational approach that is focused on continually improving the quality and efficiency with which the laboratory operates. TQM is a component of QA that supplements and increases the overall quality of the program. QA is a defined program that is focused on maximizing detection of laboratory error, while TQM aims to assist in this process by maximizing efficiency.

Unlike QA, which has no specific agenda for improvement, TQM is an approach that seeks to continually review and reevaluate the effectiveness of the QC and QA programs. If the analysis of QA data identifies a certain task as a frequent source of laboratory error, then TQM practice dictates that the error-prone task should be either modified or eliminated. Improvements are facilitated by emphasizing preventive measures that minimize the opportunity for error. Therefore, TQM is an approach that has some degree of flexibility as the agenda is to continually seek improvement in the QC and QA systems already in place. A thoughtful and well-planned QA system should not require major modifications; however, room should be allowed for minor modifications that improve the quality of laboratory operation.

Depending on the available manpower, it is most effective to assign two separate senior technical staff members: one, the responsibility of supervisor for TQM, and the other, supervisor of QA/QC. In the organizational hierarchy, TQM is not above QA, but is an adjunct. To be effective, frequent communication between the two supervisors and all staff members must be mandated.

Several examples of errors that may not be addressed by QA include:

1. Near misses (errors detected as they occur, and which are corrected before results are released). Often, these remain unreported to the QA supervisor. Such errors are almost always underreported and may indicate a flaw in the QA program. One possible reason for this is the negative connotation associated with filling out an "error report form" or a "QA incident report sheet". A technical staff member may be reluctant to point out their own error if the error was corrected immediately and no consequences ensued. Therefore, a flaw in the QA program may remain undetected and uncorrected and may potentially cause a major error in the future. One solution would be to rename the "error report form" or "QA incident report sheet". Give the report form a positive-sounding title such as "QA improvement report", and then the staff member who detects and reports the error should be rewarded for finding the flaw, thus helping to improve the QA system.
2. If the reagent refrigerator is not well organized, infrequently used kits or reagents may expire on the shelves unnoticed. A solution would be to color code all reagents with expiration dates (by year and month) as this will help identify kits and reagents to be removed from shelves.

TQM is not only concerned with the monitoring of the QC/QA program, but should also include other technical or administrative considerations that may indirectly influence the quality and efficiency of the laboratory operation.

2. Evaluation of Laboratory Staff

It is important to periodically evaluate the performance of all laboratory staff members. Individual meetings should be held with each employee to discuss aspects of work performance as well as ways to improve performance. Aspects that should be evaluated include:

- *Quality of work* — Does the employee perform the work assignments with relatively few mistakes?
- *Quantity of work* — Is work time adequately managed so that productivity is at the level expected and consistent with co-workers?
- *Interpersonal relationships* — Can the employee work effectively and cooperate with other staff members? If time permits, do they voluntarily assist in sharing other work assignments as work loads vary.
- *Safety* — Does the employee adhere to all safety regulations and practices?
- *Punctuality and absenteeism* — Can the employee be relied upon to be at work on time and with regularity?
- *Professionalism* — Does the employee maintain a professional attitude and appearance?
- *Training programs for new staff members* — Is there an organized

training rotation for new employees and are the new employees evaluated at the end of their training period and informed of ways to improve their performance?
- *Cross-training* — Are all employees cross-trained in other laboratory areas? Rotating assignments wherein all personnel take responsibility for all areas removes walls of blame when an error does occur.
- *Salaries* — Are salaries appropriate for educational background and experience and comparable to similar laboratories in the community?
- *Work shifts* — Are work shifts and assignments properly arranged for effective operation of the laboratory?

3. Continuing Education

The amount of information accumulating about AIDS and HIV testing is rapidly evolving; therefore, it is critical for laboratory staff to remain informed of current policies and changes in technical information concerning HIV testing. This can be accomplished through periodic in-service training, guest lectures, circulation of pertinent articles or newsletters, and allowing participation in other continuing education programs. A good QA program includes a mechanism that continually updates personnel so that they remain current on changes and improvements in the testing and QC fields.

4. The Laboratory as a Diagnostic System

Another way to view the application of TQM principles in the HIV testing laboratory is to consider the laboratory itself as a diagnostic system where specimens enter the system, and results emerge. With this concept, system variability becomes the primary source of error.

By examining the entire laboratory operation as a whole, it becomes apparent that human error is the primary source of system variability in the HIV diagnostic system. Human error is then further categorized as that due to either technical or system error. Technical error can be minimized by reinforcing education, training, motivational approaches, and by rechecking all aspects of laboratory performance. System error, however, is more difficult to address since it is often manifested in subtle changes that may not be readily apparent until a failure occurs. System failures, although random, infrequent, and unintentional, can result in major consequences. System error can be made to nearly approach zero by incorporating an increased degree of redundancy into the system design. Redundancy in the system refers to overlapping measures that "foolproof" the system and consequently place a much greater degree of reliability in the quality of results. A more reliable system is inherently a more efficient system. Although many methods are available to increase redundancy in the HIV testing algorithm, the following example will illustrate how the use of two different ELISA tests can increase the reliability, and thus the efficiency of the HIV screening process.

The HIV ELISA screening test used on all samples entering the system (ELISA test A) is very sensitive (99.9%), but somewhat less specific (94.5%)

and has a test efficiency of 94.7% (0.947). By using ELISA test A, virtually all true positive samples are detected, although there are a number of false-positives. The sample is now repeated in duplicate by a second technologist with a more specific test (ELISA test B), based on a different principle, and that has 96.5% sensitivity, a 99.5% specificity, and a test efficiency of 96.1% (0.961). Repeating all reactive samples from test A in duplicate by a second technologist using ELISA test B will correctly reclassify most false-positives and false-negatives. By using such a system for screening, the test efficiency "E", or probability that each sample is correctly categorized, is increased since we are using two different assays of complementary sensitivities and specificities, and a second technologist for repeat testing. The new degree of efficiency can be established by substituting the individual efficiencies of each testing situation into a common probability equation as shown:

$$\begin{aligned} E &= 1 - [1 - p(A)][1 - p(B)] \\ &= 1 - [1 - (0.947)][1 - (0.961)] \\ &= 1 - [0.053][0.039] \\ &= 1 - 0.002 \\ &= 99.8\% \text{ efficiency.} \end{aligned}$$

Therefore, two samples in 1000 will be misclassified by this dual ELISA screening system. This is also a more economical approach in the long run, considering the cost of performing a WB on a false-positive sample. *(Note: The stated sensitivities and specificities of commercially available assays are an estimate and are entirely dependent on the population selected for the determination, the skill of the operator of the equipment, and the operating conditions of the equipment. These parameters should be determined by each laboratory using the kit with their own populations.)*

To further assess the ability of redundant measures to increase testing efficiency, it is important to first differentiate the different types of false-positives and -negatives. There are biologic false-positives and -negatives due to characteristics of the sera (e.g., interfering antibodies), and technical false-positives and -negatives. The technical false-positives and -negatives may be due to a single error or a consistent error in technique (e.g., careless pipetting).

At a laboratory in which the initial and repeat testing is performed by the same technologist using the same test, nearly all false results due to random technical error (the majority of technical errors) will be correctly reclassified. However, if the error was due to a consistent error in technique, and the same technologist repeats the assay, errors may or may not be detected. If poor technique causes an error, repeating the assay in the same manner will most often simply repeat the error. Furthermore, if the error was due to a biologic false-positive or -negative (increasingly less common due to improved test characteristics), repeat testing using the same assay, regardless of who does the repeat testing, will most likely not reclassify the result. The testing algorithm

TABLE 9
Redundancy and Test Efficiency

	Repeat using same assay, same tech	Repeat using different assay, same tech	Repeat using different assay, different tech
Technical FP or FN (random error)	+	+	+
Technical FP or FN (technique error)	−	±	+
Biologic FP or FN	−	±	+

+ The measure will most likely correctly reclassify the false result.
± The measure may or may not reclassify the false result.
− The measure most likely will not reclassify the result.

most capable of maximizing detection of all classes of false results is to repeat all samples in duplicate using a test based on a different principle, and repeat the testing by a second technologist (Table 9).

Nearly all laboratories performing HIV testing have policies that require repeating reactive ELISA results in duplicate; however, the following additional redundant measures make the HIV diagnostic system more foolproof and therefore, more reliable and efficient:

- The use of two different screening assays based on two different principles (i.e., indirect vs. competitive), each of slightly different sensitivities and specificities, with the more sensitive ELISA used first.
- Establish rotating laboratory assignments in which repeat testing is always performed by a second technologist or by the laboratory supervisor.

Some QA systems have an additional requirement that all positive results be verified by a second specimen submitted from the patient, although this approach is overly burdensome for most hospital laboratories.

The cost of QA in time and resources can be as much as 25 to 30% of the cost of performing an HIV assay; yet when considering the consequences of reporting an incorrect result, the cost of QA is small.

V. QUALITY CONTROL: MONITORING THE TESTING PROCESS

As mentioned previously, QC refers to those measures that must be included during each assay in order to verify that the test is working properly. The handling of controls involves much more than just simply including them on each run. Controls must be correctly interpreted and recorded as a permanent QC record. Thus, QC indicates that each run produces results that are as accurate as the limits of the test allow.

To properly validate assay performance, the controls are run simultaneously and under the same conditions as the unknown samples. Upon completion of the assay procedure, the kit criteria are used to judge the internal controls, and then the sample results are interpreted using the same criteria. When the controls produce acceptable results, the QC indicates that the assay is valid, all test conditions for that run have been met, and all test results for that run are reliable.

QC does not, however, indicate that results are accurate. Accuracy is dependent on the characteristics and limitations of the tests as well as operator characteristics (technical error). Furthermore, QC does not indicate that the results have been reported properly, or reported on the correct patient.

The following items are essential elements of quality control that must be performed during every assay:

1. Each test run must include one full set of controls. A "run" is any number of ELISA plates or IFA slides that are used in an assay concurrently. Each subsequent plate must include at least one of each control. The values from controls that were included on the first run of the day cannot be used for subsequent runs.
2. The controls for each test run must yield results within the limits of the manufacturer's criteria for acceptability and validity of the run.
3. Any run not having at least the minimum number of controls falling within the acceptable range is invalid and MUST be repeated (see test kit insert for acceptable ranges).
4. All test kits MUST be used before the expiration date to ensure valid results.
5. Physical parameters of the test such as incubation time and temperature must be followed to ensure proper performance.

A. Internal and External Controls

Ordinarily, each HIV test kit has a set of positive and negative control samples (test controls) that are to be included in each test run. These controls included with each test kit are considered internal controls or standards, while any other controls included in the run are referred to as external controls. Fundamentals concerning controls include:

- Internal controls are essential for QC measures for each run and are intended for use only with the lot number of the corresponding test kit.
- External controls should be included on each HIV run to monitor consistent performance, lot-to-lot variation between kits, and to serve as an indicator of assay performance on samples that are borderline reactors.

- Ideally a set of external controls should include a positive control, a borderline reactor, and a negative control. However, if the laboratory chooses to use only one external control, the borderline reactor must have priority (see explanation below).

Since all test controls are included with all runs, observing the O.D. readings for each control during each run allows for monitoring of inter- and intrarun performance. The most important external control to include is a borderline reactor (near the cutoff O.D.) that brings attention to any minor changes in assay performance on samples that yield O.D. values near the cutoff. If a borderline external control can go out of range, especially in the negative direction, then certainly a test sample could do the same, thus yielding a possible false-negative result. Thus, any change in the status of a borderline control will indicate the potential for unknown samples with O.D. values near the cutoff to be falsely labeled as positive or negative. Ideally, strong positive and negative external controls should also be included. As stated by Dr. Epstein of the Food and Drug Administration, in the journal, *Transfusion*, "Efforts to achieve optimal sensitivity of the available assays through the use of independent control reagents should be part of quality assurance in medical laboratories."

External controls for the laboratory may be purchased commercially, made from pooled test kit controls, or made from pooled sera. In some testing situations, pooled test kit controls or pooled sera may be the best choice for economic considerations. If pooled human sera are used for HIV-positive or -negative controls, the sera should be prefiltered with a 0.8-µm biological filter to remove aggregated proteins or debris; this will increase the speed of sterile filtering. Next, the pool should be filtered using a 0.2-µm biological filter to remove contaminating bacteria. The laboratory that chooses to use pooled human sera may also heat inactivate the pool at 56°C for 30 min. This will reduce residual HIV infectivity to below detectable limits (although it is uncertain if all virus is destroyed).

The borderline or low positive external control can be easily produced by serially diluting (Chapter 9) a known strong positive serum, assaying the dilutions, then selecting the appropriate titer for the control (see below). Dilutions of a positive pool are also acceptable; however, the chances of picking up a nonspecific interfering antibody is increased. Normal human serum (NHS) should be used as the diluent rather than normal saline (or other serum/sample diluents, etc.) to keep the antibodies in a natural serum protein environment. Once the titer is determined, a dilution is selected that produces a value just slightly above the cutoff for an indirect ELISA, or slightly below cutoff for a competitive ELISA. The authors suggest the preparation of the dilution first, the removal of one or two aliquots, and freezing of the bulk "source" until the reproducibility of the serum is verified. The control is then thawed, mixed, and aliquoted. *[Note: (1) Do not select a dilution with a value so low that the O.D. fluctuates above and below the cutoff due to normal*

Quality Control and Quality Assurance

variations. (2) Do not select a dilution that is too strong, because a control with a high O.D. (well above the cutoff) would be of limited use for borderline monitoring. Approximately 1.5 ×C.O. for an indirect sandwich ELISA, or 0.67 × C.O. for a competitive ELISA are good guidelines.]

Once the titer is determined and the selection is made, the dilution must be tested before aliquoting. This assures that the dilution gives consistent results over 10 d or at least 10 different runs. Aliquots are then prepared and stored as:

1. Serum or plasma to be used for external controls should be in sufficient quantities to last about 1 year. This allows enough quantity for future parallel testing while preparing a new lot.
2. Sera should be homogeneous, sterile, and contain no preservatives. Sterile filters can be used to filter stock solutions before aliquoting in sterile tubes.
3. These controls should be aliquoted, labeled, and stored at –20°C in a nonself defrosting freezer for up to 1 year. The aliquots should contain sufficient volume for 1 week's testing (200 µl per vial is adequate).
4. It may be desirable to evenly split the aliquots for storage into two different freezers (in case of freezer failure).
5. Once thawed, the aliquot should be stored at 2° to 8°C and discarded at the end of the week, if not completely used.
6. It may be helpful to make a small batch of backup control as well. This will be invaluable if the need ever arises for troubleshooting problems with the external controls.

If the laboratory would like to assign a lot number and expiration date to the pool, the following suggestion may be useful: a 1-year supply of a borderline reactor control that was prepared on 7 Dec 91 could be assigned lot number BR71291 (the number corresponds to the preparation date) and given an expiration date of 7 Dec 92. Under the proper storage conditions of –20°C, the control is actually stable for longer periods. The expiration date, in this case, serves only as an indicator as to when the new batch should be prepared, and does not indicate that the control will actually deteriorate on that date.

B. How to Determine Acceptance of Control Values

Most test kit inserts indicate the range of the internal controls and define the limits of out-of-range values. However, external controls are incorporated voluntarily, and their ranges must be determined individually by each laboratory. NCCLS has issued guidelines that describe the statistical treatment of control values yielding quantitative laboratory results. However, some authorities disagree with the application of statistical analysis to HIV ELISA control values because these results are qualitative. In the view of the authors, the O.D. values produced by ELISA results are directly proportional to antibody or antigen concentration, and the qualitative result is determined by a quantitative determination of the O.D. in relation to the C.O. value. Thus, statistical treatment of control values is a reliable method for monitoring the testing process.

Several statistical values will need to be established before a QC program utilizing external controls is implemented. The short-term objective of this exercise is to monitor the external controls during each run to determine if the controls yield results that are in the range expected, and that their performance is acceptable. Once these controls and their ranges are well established, the long-term goal will be to use these controls in conjunction with the regular kit controls to more thoroughly continually monitor HIV test kit performance. To accomplish this, the limits of acceptability must be determined by using several statistical tools: the mean, the standard deviation, and the coefficient of variation.

Values for the internal (test kit) controls, the external control, and the C.O. should be monitored by plotting them on QC charts. Subtle changes in control relationships are more easily observed with a visual presentation of control values over time rather than data from a single day's run.

It is beyond the scope of this chapter to present a comprehensive review of statistics, but several basic statistical tools that may be used to determine if external control O.D. values are acceptable are presented here.

Linear graph paper is used to plot QC values over time. The ordinate (y-axis) represents O.D. units or O.D./C.O. ratios, while the abscissa (x-axis) represents either the run date or run number.

C. Calculations
1. Mean

The arithmetic mean (χ) represents the average value of a set of O.D. values and is easily calculated. Let the symbol "X" represent O.D. values from the controls, and the subscript number represent each repeat of the control. Therefore, the sum of the O.D. values may be represented as $X_1, X_2, X_3,...,X_n$, and the symbol (N) will represent the total number of O.D. values. The sum of all X values (ΣX) divided by the total number of values (N) is defined as the mean, and is represented by the formula:

$$\text{where the mean } \chi = \frac{X_1 + X_2 + X_3,..., X_n}{N} = \frac{\Sigma X}{N}$$

example: if ($X_1 = 0.675, X_2 = 0.598, X_3 = 0.702$)

$$\text{then: } \chi = \frac{0.675 = 0.598 + 0.702}{3} = 0.658$$

Therefore, in this example, the mean of the external controls after three runs is 0.658. We will assume for QC purposes that the degree to which the numerical data tend to disperse about the mean can be represented as a bell-shaped curve called a Gaussian, or normal, curve (Figure 43). The curve gets its bell shape from the fact that more values are found near the mean and increasingly fewer values are found on either side away from the mean.

To be statistically significant, the mean value of an external control should be obtained after testing the control on at least 15 to 30 runs. Strong positive controls are usually much more variable than borderline or negative controls; this is due in part to the differences in their relative ranges. For example, the range of an internal positive control given by the package insert will often be given as any value equal to or above a certain O.D. (i.e., the positive control must be ≥0.800 to be valid). This translates to a range of 0.800 to ≥2.000 for the internal positive control, while the ranges for negative and borderline reactors will necessarily be much more narrow. Thus, strong positive controls may require 100 or more values to establish a reliable mean. In general, the greater the differences between particular O.D. values and the mean (referred to as deviations from the mean), the greater the number of values required to obtain a reliable control mean.

2. Standard Deviation

The standard deviation (s) is a measure of the variation or dispersion about the mean and defines the expected range of a control in relation to the mean value. The standard deviation of a set of O.D. values can be represented as "s" and is defined by the formula:

$$s = \sqrt{\frac{\Sigma(X_n - \chi)^2}{N}}$$

To establish an acceptable range for external controls, first determine the mean value of the control and then calculate the standard deviation. The numeric value obtained from this formula represents one standard deviation. The mean value plus or minus one standard deviation establishes a range around the mean in which approximately 68% of all values would be expected to fall. Similarly, the mean plus or minus two standard deviations and three standard deviations represents the range where 95 and 99% of all values would be found, respectively. Figure 44 represents the approximate normal distribution of an external control value that can be expected at different standard deviations.

Specifying an interval around the population mean within which a control value may be expected to fall with some known degree of confidence is what is commonly referred to as a confidence interval. The confidence interval of one standard deviation is approximately 68%. This means that on any given run an external control would be expected to fall in this range 68% of the time, while 32% of the time it would fall outside of this range due to random error (or chance) alone, even if the run is performed perfectly. Any control outside

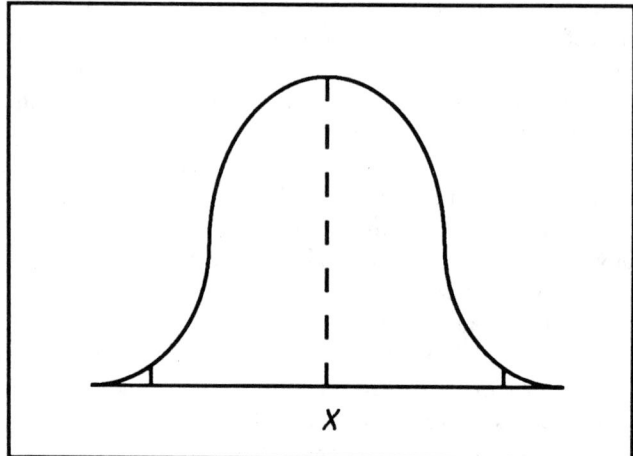

FIGURE 43. Gaussian curve.

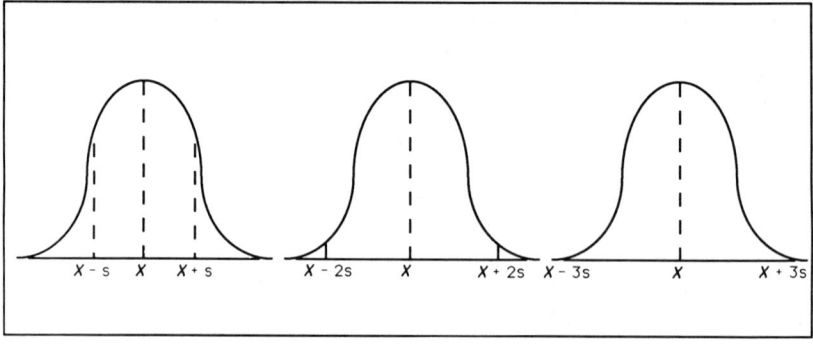

FIGURE 44. Approximate normal distribution.

one standard deviation, therefore, may or may not represent a significant QC problem, since 32% of the time it is random error. However, 68% of the time it would be due to a problem, or a significant error. Keep in mind that if the external control range selected is based on one standard deviation, an external control falling outside the range of one standard deviation would have to be rejected, even though 32% of the time the value would be due to a random error (chance alone and a problem-free assay). The selection of the appropriate confidence interval for external controls is discussed further in Section V.G.

3. Coefficient of Variation*

The coefficient of variation (C.V.) is determined by dividing the standard

---

* Two methods of calculating control ranges (O.D. and O.D./C.O.) are described in Section D.

deviation by the mean, and then multiplying by 100. The value obtained is a reflection of the relative dispersion of O.D. values around the mean, and is useful as an indicator of reproducibility of control values or samples that are repeated in duplicate. The formula below is unitless and is expressed as a percent.

$$\text{C.V.} = \frac{s}{\chi} \times 100$$

A consistently high C.V. (≥20%) between replicates of kit controls is an indication that pipetting is not consistent, or possibly that the pipette requires recalibration. This can occur, for example, when the same tip is used for all replicates of a negative or a positive control. Changing tips each time, even when pipetteting the same control in more than one well, will greatly reduce this variation, and yield more consistent performance of controls (see Chapter 9, Section IV).

The standard deviation, and thus the C.V., are affected by aberrant O.D. values. When determining a range for a given control, the reliability of the mean, the standard deviation, and the C.V. increase as the number of determinations increase. An outlying value would have a much greater effect on a set of 15 values than on a set of 35 values.

A control range is selected based on the number of standard deviations around the mean; this determination is a decision made by the individual laboratory, depending on the confidence interval desired. Once a range is determined, any run with a control value that falls outside of this range must be rejected and the run repeated; this is what is known as a rejection or **control rule**. A C.V. of ≤20% indicates that the controls perform consistently, with relatively little dispersion about the mean, which indicates the control ranges based on the standard deviation should be reliable (see Section V.E).

To illustrate the use of these three formulae, the mean (χ), the standard deviation (s), and coefficient of variation (C.V.) are calculated, given the hypothetical low positive (borderline) external control O.D. values, performed on 25 different occasions, as indicated in Table 10.

To calculate the standard deviation of the low positive control mean, first find the absolute value of each given O.D. value (X_n) minus the mean (χ) of all values and set up a table as indicated in Table 11. The mean of column 1 (χ) = 0.591. Next, find the sum of column 4, which equals the sum of the squares of the absolute values:

$$\Sigma(X_n - \chi)^2 = 0.355$$

N = total number of values (a total of 25 exist), thus, the standard deviation of these control values is calculated as follows:

Calculate the range based on the standard deviation:

TABLE 10
Raw O.D. and C.O. Values for an External Control

	External control O.D. value	Cutoff value (C.O.)
X_1	0.598	0.293
X_2	0.694	0.292
X_3	0.556	0.245
X_4	0.615	0.208
X_5	0.432	0.198
X_6	0.707	0.310
X_7	0.857	0.282
X_8	0.686	0.274
X_9	0.512	0.202
X_{10}	0.535	0.340
X_{11}	0.656	0.269
X_{12}	0.634	0.209
X_{13}	0.553	0.235
X_{14}	0.484	0.198
X_{15}	0.540	0.297
X_{16}	0.659	0.273
X_{17}	0.856	0.302
X_{18}	0.508	0.240
X_{19}	0.444	0.314
X_{20}	0.376	0.268
X_{21}	0.492	0.222
X_{22}	0.634	0.274
X_{23}	0.582	0.301
X_{24}	0.520	0.259
X_{25}	0.638	0.181

Note: We will refer to this set of data again to illustrate several graphing alternatives.

$$s = \sqrt{\frac{\Sigma(X - \chi)^2}{N}}$$

$$s = \sqrt{\frac{0.350}{25}} = 0.119$$

s	Ranges ($\chi \pm s$)
1s = 0.119	(0.472–0.710)
2s = 0.238	(0.353–0.829)
3s = 0.357	(0.234–0.948)

TABLE 11
Standard Deviation Worksheet

Column: 1		2	3	4
	Low positive external control O.D.	$\|X_n - \chi\|$	Absolute value[a] (d)	$\|X_n - \chi\|^{2}$ [b] (d^2)
X_1	0.598	0.598 – 0.591	0.007	0.000[c]
X_2	0.694	0.694 – 0.591	0.103	0.011
X_3	0.556	0.556 – 0.591	0.035	0.001
X_4	0.615	0.615 – 0.591	0.024	0.001
X_5	0.432	0.432 – 0.591	0.159	0.025
X_6	0.707	0.707 – 0.591	0.116	0.013
X_7	0.857	0.857 – 0.591	0.266	0.071
X_8	0.686	0.686 – 0.591	0.095	0.009
X_9	0.512	0.512 – 0.591	0.079	0.006
X_{10}	0.535	0.535 – 0.591	0.056	0.003
X_{11}	0.656	0.656 – 0.591	0.065	0.004
X_{12}	0.634	0.634 – 0.591	0.043	0.002
X_{13}	0.553	0.553 – 0.591	0.038	0.001
X_{14}	0.484	0.484 – 0.591	0.107	0.011
X_{15}	0.540	0.540 – 0.591	0.051	0.003
X_{16}	0.659	0.659 – 0.591	0.068	0.005
X_{17}	0.856	0.856 – 0.591	0.265	0.070
X_{18}	0.508	0.508 – 0.591	0.083	0.007
X_{19}	0.444	0.444 – 0.591	0.147	0.022
X_{20}	0.376	0.376 – 0.591	0.215	0.046
X_{21}	0.492	0.482 – 0.591	0.109	0.012
X_{22}	0.634	0.634 – 0.591	0.043	0.002
X_{23}	0.582	0.582 – 0.591	0.009	0.000
X_{24}	0.520	0.520 – 0.591	0.071	0.005
X_{25}	0.638	0.638 – 0.591	0.047	0.002
		Sum of column	4 =	0.355

[a] Difference of column 2.
[b] Square of column 3.
[c] Values <0.0005 are rounded to 0.000.

D. Plotting Quality Control Graphs

A QC graph is relatively simple to formulate and is easy to maintain. Two methods may be used. The first (described below) is acceptable, but not preferred. It is represented here for the purpose of illustration and comparison.

The absorbance or O.D. value obtained from each external control is plotted on the y- or vertical axis. The x-axis is used to indicate the date of the run, or if the laboratory does more than one run per day, the x-axis can represent the run number.

If values are plotted as run number, this number must be indicated on the worksheet. All graphs shown here are adaptations of what is commonly known as a "Levy-Jennings" or "Shewhart" type chart. Figure 45 shows a simple QC graph (O.D. vs. run) indicating the low positive control and C.O. values given previously, plotted with each run. A similar chart can be adapted to show all internal and external controls, and the cutoff, allowing for a visualization of all test controls simultaneously.

Although illustrating O.D. vs. run is acceptable, a second and more accurate alternative is to plot the O.D./C.O. ratio instead of the raw O.D. values on the y-axis. This is the preferred method because the control values are expressed relative to the C.O. value (i.e., C.O.values will change slightly between runs; therefore, controls should be compared to each calculated cutoff). Using the 25 runs from the previous example, including their corresponding C.O.values, a similar table is set up as before, using ratios instead of raw O.D. values. A mean and standard deviation are calculated from the O.D./C.O. ratio (see below). The O.D./C.O. can be graphed as noted in Figure 46. Notice the change in values representing the y-axis; the range can be from 0.000 to about 6.000, depending on the type of control used (negative, low, or high positive).

The calculation for the mean, standard deviation, and control ranges are basically as before with the exception that the deviation is based on the O.D./C.O. ratio instead of the raw O.D. values. Set up a table identical to Table 12.

1. The mean of column 3(c) = 2.321.
2. To calculate the standard deviation, find the absolute value derived from the difference of each ratio (column 4). Then find the sum of the squares of the absolute values (column 5).

$$\Sigma(X_n - \chi)^2 = 6.141$$

Finally, calculate the standard deviation.

$$s = \sqrt{\frac{6.141}{25}} = 0.496$$

The ranges based on the standard deviation are as follows (graphed in Figure 46):

s	Ranges ($\chi \pm s$)
1s = 0.496	(1.825–2.817)
2s = 0.992	(1.329–3.313)
3s = 1.488	(0.833–3.809)

Notice that the O.D. values shown as two lines on Figure 45 are now

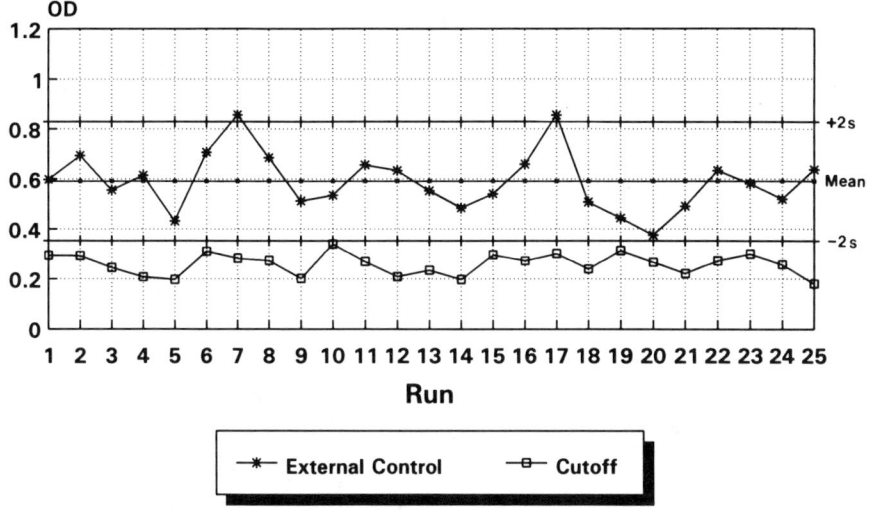

FIGURE 45. A simple QC chart.

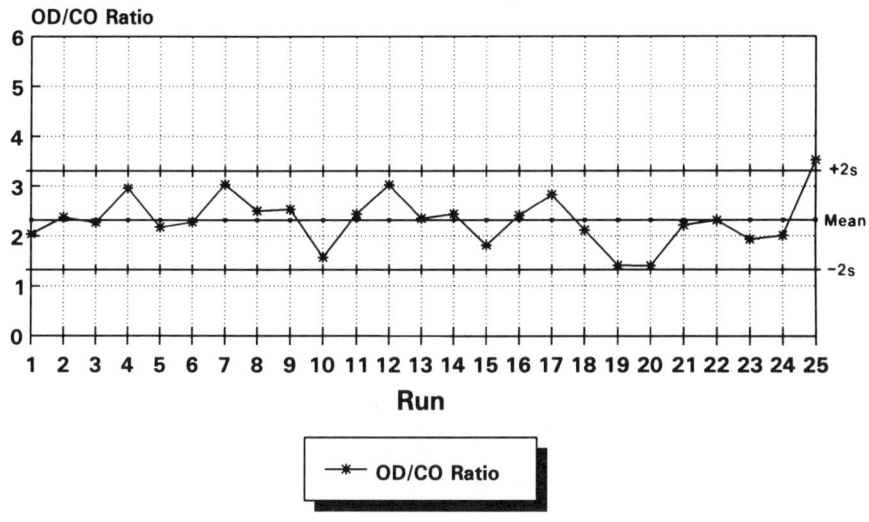

FIGURE 46. Illustration of O.D./C.O. vs. run.

combined and shown as one line on Figure 46. This technique can also be used to graph the high positive and negative values.

E. Use of Calculated Values

Exercise 1: Given the ranges calculated from Table 11, plot the external low positive control O.D. values from the following six runs on a QC graph. Then determine which of the six values are acceptable using 2s (two standard deviations) as limits.

TABLE 12
O.D./C.O. Values

	Column: 1	2	3	4	5
	External control O.D. value	Cutoff value (C.O.)	O.D./C.O. ratio	d^a	$d^{2\ b}$
X_1	0.598	0.293	2.041	0.280	0.078
X_2	0.694	0.292	2.377	0.056	0.003
X_3	0.556	0.245	2.269	0.052	0.002
X_4	0.615	0.208	2.957	0.636	0.404
X_5	0.432	0.198	2.182	0.139	0.019
X_6	0.707	0.310	2.281	0.040	0.002
X_7	0.857	0.282	3.039	0.718	0.516
X_8	0.686	0.274	2.504	0.183	0.033
X_9	0.512	0.202	2.535	0.214	0.046
X_{10}	0.535	0.340	1.574	0.747	0.558
X_{11}	0.656	0.269	2.439	0.118	0.014
X_{12}	0.634	0.209	3.033	0.712	0.507
X_{13}	0.553	0.235	2.353	0.032	0.001
X_{14}	0.484	0.198	2.444	0.123	0.015
X_{15}	0.540	0.297	1.818	0.503	0.253
X_{16}	0.659	0.273	2.414	0.093	0.009
X_{17}	0.856	0.302	2.834	0.513	0.263
X_{18}	0.508	0.240	2.117	0.204	0.042
X_{19}	0.444	0.314	1.414	0.907	0.823
X_{20}	0.376	0.268	1.403	0.918	0.843
X_{21}	0.492	0.222	2.216	0.105	0.011
X_{22}	0.634	0.274	2.314	0.007	0.000^c
X_{23}	0.582	0.301	1.934	0.387	0.150
X_{24}	0.520	0.259	2.008	0.313	0.098
X_{25}	0.638	0.181	3.525	1.204	1.450
Sum	=	6.141			

[a] "d" (column 4) equals the difference of each value in column 3 minus the mean of column 3, or $|X_n - \chi|$.
[b] Column 5 equals the square of column 4, or $(|X_n - \chi|)^2$.
[c] Values <0.0005 are rounded to 0.000.

s	Ranges ($\chi \pm s$)
1s = 0.119	(0.472–0.710)
2s = 0.238	(0.353–0.829)
3s = 0.357	(0.234–0.948)

Run no.	O.D. value	C.O. value
1	0.340	0.239

Quality Control and Quality Assurance

2	0.780	0.366
3	0.956	0.326
4	0.526	0.261
5	0.813	0.211
6	0.473	0.239

Solution: Plot a simple graph using these raw values, as before. The simple plot would appear similar to Figure 47, and, as noted in the figure, it is readily apparent that the range selected affects whether the external control value is acceptable. By choosing a one standard deviation range, runs 4 and 6 show the only acceptable control values. Based on two standard deviations, runs 2, 4, 5, and 6 are acceptable. By increasing the limit to three standard deviations, all values are in range, except the third. Therefore, if two standard deviations are chosen, runs 1 and 3 would be rejected because the external controls did not fall within acceptable limits. As more standard deviations are selected, more values will be considered acceptable; however, a greater chance exists that these values were due to significant errors.

Exercise 2: Using the six run values from above, set up a graph of O.D/C.O. ratios using the O.D./C.O ranges from Table 12:

s	Ranges ($\chi \pm s$)
1s = 0.496	(1.825–2.817)
2s = 0.992	(1.329–3.313)
3s = 1.488	(0.833–3.809)

Solution: Find the O.D./C.O. ratio as before.

Run no.	O.D. value	C.O. value	O.D/ C.O.
1	0.340	0.239	1.421
2	0.780	0.366	2.130
3	0.956	0.326	2.924
4	0.526	0.261	2.013
5	0.813	0.211	3.856
6	0.473	0.239	1.982

Set up a graph similar to Figure 45, chart the six values, and observe which of the O.D./C.O. values are in range. The QC graph should be similar to that depicted in Figure 48.

By comparing the six run external control results via two different graphing methods (Figure 49), it is evident that the same external control may be accepted using one method, but not accepted using the other. By plotting raw

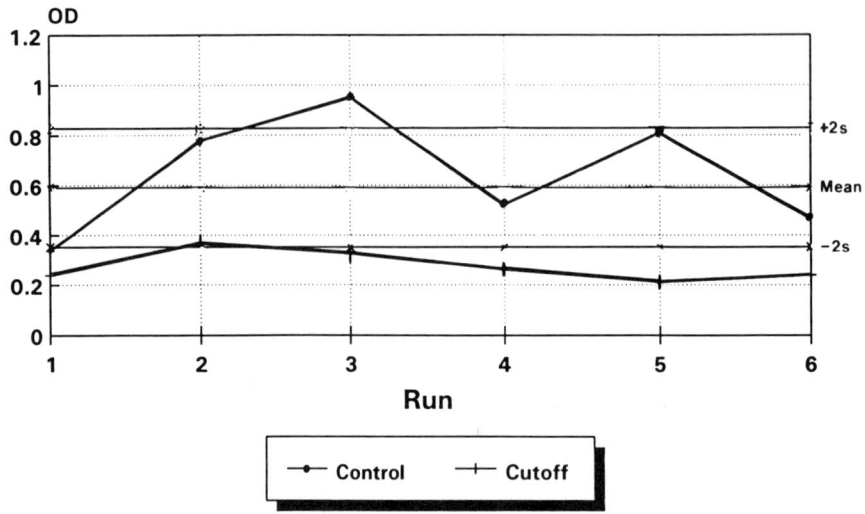

FIGURE 47. Simple graph (Exercise 1).

O.D. values alone, the control in runs 1 and 3 is rejected, while the control in run 5 is accepted (using either two or three standard deviations as limits).

In contrast, when plotting O.D./C.O. values (using the same 2s limit), the control in runs 1 and 3 now become acceptable, while the control in run 5 is now unacceptable. As noted on run 3, the control value was outside of three standard deviations when plotting simple O.D. values, and therefore was rejected. However, when the same value is compared in relation to the cutoff (O.D/C.O.), its value then becomes acceptable within two standard deviations because the cutoff on that run was very high. Likewise, on run 5 the control fell inside of two standard deviations by the simple graph, yet was rejected when plotting O.D./C.O. because the cutoff value was very low on this run.

This hypothetical set of values illustrates two QC graphing alternatives. Both methods can be effective QC tools, although the second method can give a more accurate assessment by taking into consideration daily fluctuations in O.D. values. It may take several months to establish ranges for external controls, and it does take some time to initiate the use of these control systems. However, this type of control system can be very effective in helping to identify potential problems and inaccuracies in the laboratory setting.

F. Shifts and Trends

Shifts occur when control values of six consecutive runs fall on one side of the mean line. This usually indicates that a major change has occurred. Recall that a normal distribution should fall randomly above and below the mean. Examples of situations that can cause shifts include:

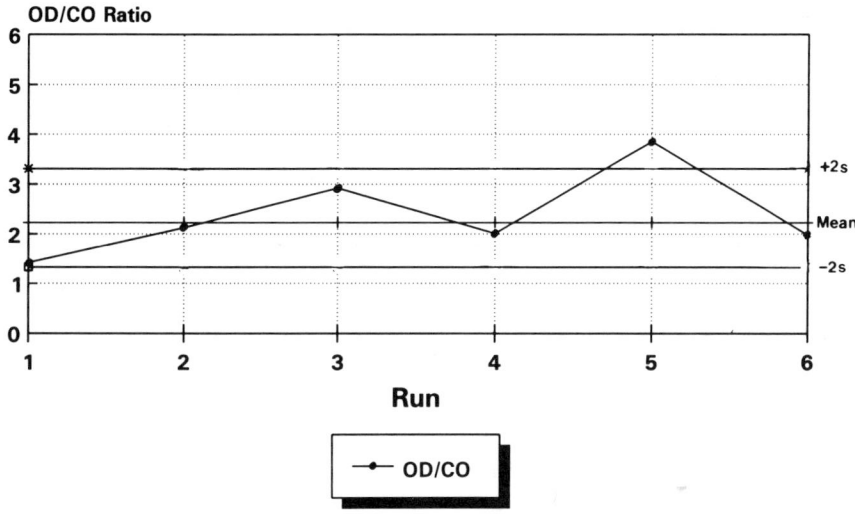

FIGURE 48. O.D./C.O graph (Exercise 2).

- Switching to a new lot of kits
- New reagents, including fresh wash buffer
- Changes in incubation temperatures (perhaps the water bath was not turned on, or the temperature was changed)
- A new technician (who may have different pipetteting techniques)
- Changes of equipment (pipettes, plate washer, etc.)

Trends occur when six successive points become distributed in one general direction. This is usually the result of slowly changing parameters such as deterioration of reagents and faltering equipment (e.g., a routinely used pipette slowly losing its calibration).

The most common reasons for shifts and trends are deteriorating reagents or kit controls. Conjugates are generally the first reagents to deteriorate as a kit ages. Figure 50 illustrates the concepts of shifts and trends.

Sometimes a manufacturer produces reagent lots that pass their QC requirements, but fail to perform properly in the field. A contributing factor may be shipping or storage conditions. These reagents must be identified as quickly as possible. A comparison of performance of new lots of kits to the previous lots via a common control material (external controls, or a well-characterized panel of sera of various reactivities) is essential for this purpose. This comparison is called "parallel testing" and can be accomplished by including an external control along with the controls from the previous lot, on the first run of the new lot. If all controls produce the expected results, then the new lot has passed the parallel test, and the new lot of kits is ready for routine use. A parallel test is

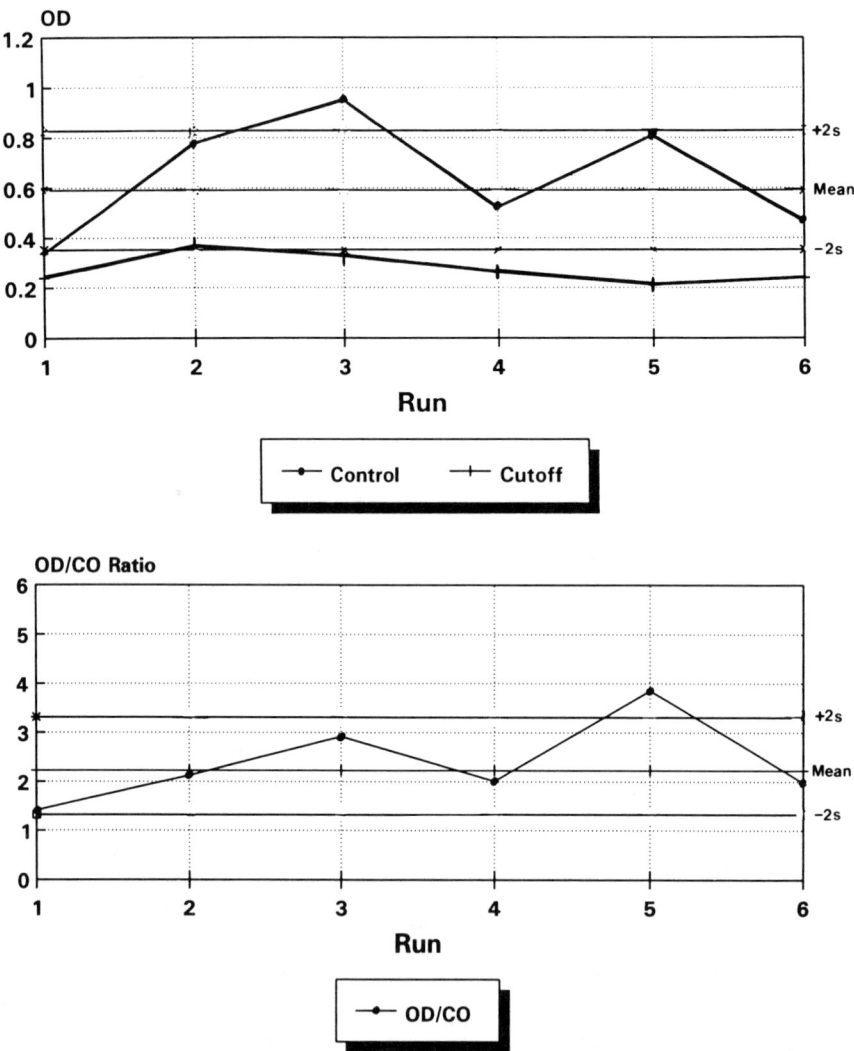

FIGURE 49. Comparison between Figures 47 and 48.

the only valid reason for using a set of controls on a lot other than the one for which it was intended. Some provisos follow:

1. When a new lot of test kits is introduced into the testing process the date should be noted on the control chart. Be especially vigilant against sudden shifts in all control values. Some variation, either up or down, is expected. Any misclassification of controls or extreme shifts in values should be recorded and the manufacturer notified. The acceptable degree of variation must be determined by each laboratory (see below).

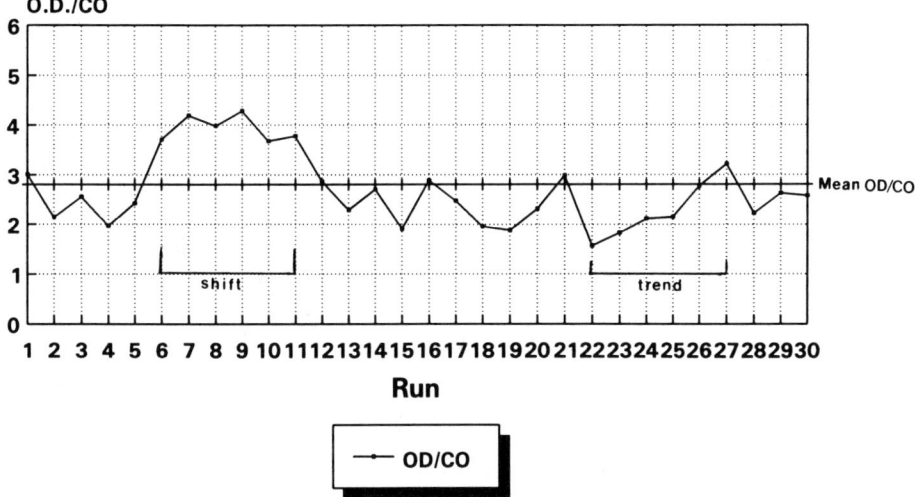

FIGURE 50. Shifts (6 to 11) and trends (22 to 27).

2. Monitor the performance of the external positive control vs. the kit positive control. Internal kit controls are adjusted by the manufacturer so that an expected range of values is obtained with each lot of kits. The authors have used kits in which the internal kit controls give the same value as the previous lot, but the external controls show a significant shift in values. This is most likely due to the artificially stabilized kit controls. Differences, however, should be minimal. Again, the acceptable degree of difference must be determined by each laboratory.
3. Monitor the external and internal negative controls for trends toward higher or lower values. Such a change may suggest reagent problems but may also indicate technique or equipment problems (e.g., washer out of alignment or needs cleaning).
4. Review the average absorbance value for the specimens yielding negative results. An increase in this value may indicate problems with technique, equipment, or reagents. Similarly, increases in the number of initially reactive specimens that do not repeat or that do not confirm by WB are signals of the same types of problem.

G. An Example of a Quality Control Protocol

A written QC policy in the laboratory SOP manual that defines the limits of acceptability of control values can be called a rejection or control rule. A control rule, such as a range of one standard deviation about the control mean, is designed to detect errors in the assay. However, a single control rule in itself cannot sufficiently detect all error signals without producing some rate of false rejection due to random error. For this reason, it is best to have a combination of rules that maximize error detection while minimizing false rejection of a run. The following is a set of control rules that may be used in a clinical laboratory that is based on the inclusion of a borderline external control in duplicate on each run. A run is closely scrutinized and the final decision to reject the run is made when:

- Both external control results exceed two standard deviations from the mean
- One of the external controls exceeds three standard deviations
- One external control exceeds two standard deviations on two consecutive days
- If evidence of a shift or trend becomes apparent (6 data points from one duplicate or 12 data points from duplicates are in one direction or are on one side of the mean)

A control rule of 2 standard deviations by itself is too strict a policy for an external control since approximately 1 run in 22 would be rejected due to random error alone. A rule such as this using two controls would result in 1 run in 11 ($1/22 \times 1/2 = 1/11$) being rejected.

The authors recommend a combination of rules such as above, including a range of between 2 to 2.5s above and below the mean for evaluation of external controls (standard deviation does not necessarily have to be expressed as a whole number). Thus, by adopting a combination of rules including the 2 to 2.5s limit, any control that falls within this range will have the level of confidence of at least 95%; only 5% or less of any set of control values will fall outside of this range due to chance (random error). This means that when compared to a control outside one standard deviation, such a value outside two standard deviations has a greater chance (95 vs. 68%) of representing a significant QC problem rather than random error alone. Adopting a rule of three standard deviations would eliminate most rejection due to random error, but the chances of detecting a true error rejection signal would be significantly diminished.

The final decision to reject the run does not rely on hard and fast rejection rules alone. The rejection rules are "flags" that indicate that all aspects of the run must be carefully considered before results of that run are released. The laboratory manager must have the final authority to release results from a run (even if some controls are out), as long as it is determined that the overall quality of the results are still reliable. This position is supported by the NCCLS (NCCLS Document C24-A, 11, 6, 1991).

H. Use of Duplicate Controls

Another method to monitor quality using internal and external controls is to repeat one control (either internal or external) in triplicate on the same run and compare the results; this will show "intrarun reproducibility". Differences

Quality Control and Quality Assurance 153

usually indicate pipetteting errors, but may indicate equipment failure (see Section IX). Also, a few controls could be run on 3 consecutive days to evaluate "interrun reproducibility" (occurrence of shifts and trends). Via either of these methods, the C.V. should not exceed 20%.

Adding a single borderline control to each run ensures that the laboratory may more accurately monitor consistency in both run-to-run and lot-to-lot test performance. Now consider the consequences of adding this single control at the beginning of the plate, immediately after the internal controls. The information from this control would indeed indicate the overall run performance and run-to-run consistency, yet it would not give any indication of the variation in performance from samples at the beginning of the plate in comparison to the samples at the end of the plate. For this reason the authors suggest running the borderline external controls in duplicate, one control immediately after the internal controls at the beginning of the plate and the second in the last well at the end of the plate. This establishes a second QC check, verifying that samples are performing properly throughout the plate. For best results, the technologist should not add the external control to both wells while the vial is still open, as this defeats the purpose of running the external control in duplicate. These extra measures can greatly increase the chances of identifying such systematic error as differences in reagent volume from the first row to the last row if using multichannel pipettes for dispensing. For example, when using an eight-channel pipette to dispense conjugate to a 96-well plate, if one does not aspirate and eject once before proceeding (prewetting), the delivery volume to row 1 (containing internal controls) will be slightly less than rows 2 through 12. This type of technical error will slightly increase the O.D. in the first row that contains controls in relation to the subsequent rows, possibly causing a borderline or low positive reactor to become a false-negative (see Chapter 9, Section IV). Other examples include:

- Plate washers not washing consistently
- Conjugate or substrate deterioration between early and late additions of samples to the plate
- Contamination from any number of sources
- Differences in incubation times between early and late wells

I. Recalculation of Acceptable Control Values

Internal kit controls are optimized for reactivity and QC tested by the manufacturer before test kits are placed on the market. Under most circumstances, if all test conditions are met, licensed test kits perform quite well. As mentioned earlier, monitoring by external controls will help distinguish subtle signs that indicate that the test kit may not be performing optimally. Establishing ranges and rejection rules for a QA program that includes external controls will take some effort before initiation; however, results of the highest quality are dependent on a "foolproof" system. The approach used in establishing such a system should be to

- Calculate the mean and standard deviation of the external controls after obtaining values for 15 to 30 runs, then
- Observe the performance of the external controls through a period of at least three consecutive kit lot numbers. At the end of this period,
- Recalculate the cumulative mean and standard deviation over the entire period. After careful reevaluation, it is now possible to
- Establish rules for run rejection based on performance of external controls

During the data gathering process, the external control values obtained should not be used to decide the acceptability of a run since the recalculated ranges and the final run rejection rules have yet to be established. Initially, the laboratory must rely on the manufacturer's criteria for run acceptability, which is based on the values obtained from the internal controls (kit controls).

1. Changes in the External Control Lot

When it becomes necessary to prepare a new batch of external controls, be sure to follow the same method of preparation and select a titer similar to the previous batch. By doing this, the performance of the new batch of external control should be very close to the previous batch. Store the newly prepared batch between 2° and 8°C for a few days while an aliquot is parallel tested with the old external control. Parallel testing with the previous batch of controls should be performed before the new lot is aliquoted and frozen. The newly prepared external control will usually require some minor adjustment in titer so that the new O.D. value falls close to the old mean (+/-1s). After adjustment is complete, aliquot, label, and store the new lot of external control at −20°C, as was done previously. Before the new lot is actually instituted, use thawed aliquots in parallel with the previous lot for several weeks and preferably through one lot change of test kits. This will verify that the control was prepared properly and performs adequately after freezing and thawing. After 15 to 30 runs, calculate a new cumulative mean and standard deviation based on both the new and the old controls. **Do not change the external control lot and test kit lot simultaneously.**

2. Changes in Test Kit Lot

When monitoring performance by control charts during a kit lot change, a small shift in internal control values is usually demonstrated on the first day that the new lot is run. A shift of this type is readily explained, is usually within 1s, and requires no action. When observing the external control values, a somewhat larger shift is often observed. This shift could be significant enough to cause a run rejection if any of the control rules are broken.

The larger shift in the external control values, again, is most likely due to the artificial adjustment of the conjugate in relation to the internal controls in the new lot. This adjustment by the manufacturer is necessary to ensure consistent run-to-run as well as lot-to-lot performance of the kit in reference to internal controls; this may cause confusion to those wanting to use external controls.

The problem of out-of-range external control values during lot change can be significantly reduced if the mean and standard deviation are recalculated cumulatively after a minimum of at least three different lots of test kits are used. In fact, cumulative values can be recalculated after every lot to yield a more reliable mean and standard deviation. Thus, any run rejection due to an out-of-range control value will more likely be attributed to true systematic error. However, if a dramatic shift ($\geq 2s$) is noted in the external control value when changing lots, the manufacturer should be notified, and the problem resolved. It is possible that the new lot has been damaged by exposure to adverse conditions during shipment, or perhaps the manufacturer's QC program has failed. If a second external control is available (and has been previously characterized) it can be run in parallel to determine if the change is significant enough to affect both external controls.

J. Calculation of "Gray-Zone" Reactors

The ideal HIV screening test should have only two possible results, either reactive or nonreactive. Unfortunately, problematic samples invariably occur during routine testing. For example, some samples technically classified as negative may have an elevated O.D. reading, suggestive of some antibody activity. During early seroconversion only slightly elevated O.D. readings may be found; all samples from infected individuals must pass through a "gray zone" at some point. The gray zone is a range just below the C.O. value, where samples are not considered positive, yet the O.D. is high enough to cause suspicion.

A value of 10% is most often calculated below the cutoff, and all samples having O.D. values falling within this range are repeated or further tested using other screening tests or confirmatory assays. The calculation of 10% for the gray zone may be somewhat arbitrary, but a reasonable chance exists of finding some early HIV infections (seroconversion) by using this approach. Laboratories with "high risk" populations may wish to consider a gray zone of 20%, while those with low risk populations would be adequately vigilant in maintaining a 10% value.

K. Use of Gray-Zone Results

By adopting a policy in which gray zone reactors are investigated, an increase in samples may occur, producing indeterminate HIV results by confirmatory assays. However, these persons should be identified so that they can be informed and monitored for seroprogression. In the case of blood banking, they are prevented from donating blood. One possible disadvantage is that many individuals (perhaps 20%) with negative screening test results will produce some bands by WB; some of these individuals may produce gray zone results but are not infected.

Repeatedly gray zone-reactive specimens should be tested by WB similar to all reactive samples. If the WB is negative, then the report is sent out as negative for antibodies to HIV, with no need to mention the gray zone result. However, a gray zone reactor that is later confirmed is a sample that would have been classified as a false-negative by the screening assay. This extra measure of assurance will help to protect the blood supply, even if the threat is small.

VI. QUALITY ASSESSMENT: MONITORING LABORATORY PERFORMANCE

Quality assessment is a means to determine the quality of results. It is usually an external evaluation of a laboratory's performance that relies on incorporating proficiency panels of well-characterized sera into the testing routine. Quality assessment challenges the effectiveness of the quality assurance program. A good QC/QA program may be a substitute for quality assessment in some situations; however, quality assessment is never a substitute for good QC/QA.

For a laboratory to be respected as a testing facility, it must be a laboratory that produces excellent results, without error. If good quality control and quality assurance are diligently practiced, the laboratory will be able to meet the needs and expectations of the medical community by producing highly reliable results.

A. Proficiency Panels and Blind Testing

Several agencies operate proficiency testing programs and services for laboratories interested in quality assessment. These include WHO, the College of American Pathologists (CAP), CDC, and the American Association of Blood Banks (AABB). These and other agencies provide panels of known HIV-positive and -negative sera to participating laboratories for HIV proficiency testing. The CDC Model Performance Evaluation Program Enrollment Form is included in Appendix C.

Participating laboratories receive the panels on a regular basis (two to four times per year, depending on the program). The panels consist of a set of sera, a report form, and a return envelope. Panels should be processed and tested in a timely manner so that results can be analyzed along with those from other participating facilities. Proficiency panels should be handled routinely and objectively, and in exactly the same manner as routine samples for HIV testing. The purpose of these panels is to identify problems within a QC or QA system. Special treatment would remove this benefit. *[Note: As a rule, "proficiency testing" programs report an individual laboratory's performance scores back to them (they are often used for licensure/certification programs). "Performance evaluation", on the other hand, attempts to use aggregate data to analyze the testing process (individual reports or scores are not returned to the participants). For this reason, "performance evaluations" are not used for licensure/certification.]*

A summary analysis is prepared, and a comprehensive report is distributed to all participants. The comprehensive report will not identify any specific laboratory by name or location; however, it does include the general type of laboratory (i.e., hospital, private physician, or research laboratory, etc.) and the results obtained by each type. In addition, a confidential report of the individual laboratory's performance may be provided by some agencies. With a copy of these reports, the laboratory manager can then assess the adequacy of their QC/QA programs in their own laboratory, as well as compare their results with results obtained by similar laboratories. If a problem is identified, appropriate action can be taken.

Participation in QA programs requires relatively little time and few resources from the laboratory (inclusion of the panel in the test run); the burden of analysis of the data is borne by the agency providing the panel. The increase in the laboratory work load is offset by the value obtained from improved laboratory performance. The organizations listed below can provide information regarding proficiency programs:

WHO Anti HIV Quality Assessment Scheme
Global Program on AIDS
World Health Organization
20, Avenue Appia
CH-1211 Geneva 27 Switzerland

College of American Pathologists (CAP)
325 Waukegan Road
Northfield, IL 60093-0800 USA

American Association of Blood Banks (AABB)
1117 N. 19th Street, Suite 600
Arlington, VA 22209 USA

CDC Model Performance Evaluation Program
MPEP Survey Coordinator
Program Resources, Inc.
Science & Technology Center
P.O. Box 12794
Research Triangle Park, NC 27709 USA

A second alternative for self-evaluation is to create a panel of sera for "blind testing" (internal quality assessment). This panel should include several low positive or borderline samples, and should be submitted periodically by the supervisor or laboratory manager along with routine test samples. Certainly, the clear positive and negative samples should pose no diagnostic difficulty for the laboratory; however, by including borderline samples in a panel, a more effective challenge to quality control is assured by verifying consistency of assay performance.

The purpose of blind and proficiency testing is to evaluate the effectiveness of a QC/QA program. Participation in external proficiency programs and incorporation of blind testing panels, and comprehensive QC measures in the laboratory are elements of a good overall QA program.

VII. EQUIPMENT MAINTENANCE AND CALIBRATION

Performing regularly scheduled preventive maintenance and calibration on laboratory equipment ensures that the equipment operates optimally, the instrument may be relied on for future runs, and the performance or data produced by the instrument is the best that can be obtained from that piece of equipment. This eliminates another source of error in the diagnostic system. Preventive maintenance schedules should be included in the operating instructions of laboratory instruments such as ELISA washers and readers. Maintenance and calibration procedures should be followed regularly for optimal performance of equipment. Performance of scheduled maintenance activities should be documented for future reference.

A record or check sheet of such activities can be adopted for any given laboratory to help the laboratory manager keep track of necessary tasks. Of course, these activities can vary greatly depending on such factors as workload, types of instrumentation, how often the instrument is used, and sensitivity of the instrument to disturbance. Other factors include the number of employees, and environmental conditions surrounding the instrument such as humidity, dust, excessive vibration, or excessive heat.

All personnel handling an instrument should be properly trained in the use and care of the instrument before being allowed to operate it. All electrical connections should be routed through a surge protector to avoid potentially damaging voltage spikes.

Ordinarily, maintenance schedules are set up as tasks to perform daily, weekly, monthly, biannually, or annually. Examples of maintenance and calibration schedules with activities that can be performed are listed below. Some instruments have a built-in performance verification function that should be carried out as recommended by the manufacturer. The maintenance activities listed below are only examples of different types of maintenance tasks; individualized maintenance schedules should be set up for each laboratory, depending on the equipment utilized.

Daily: Add controls or calibrators to each run, recalibrate instruments if necessary; for ELISA readers, check that the proper filter is in place; for washers, empty waste containers, rinse sample ports with distilled H_2O or acceptable cleaners (see operator manual).

Quality Control and Quality Assurance

	Clean up all spills, check reagent levels; dispose of biohazardous waste. Check and record all temperatures (including room).
Weekly:	Keep optical and other components free from dust; clean surfaces of instruments, prepare fresh batches of reagents as needed, recalibrate instruments if necessary.
Monthly:	Perform electronic or optical checks on all components. Many automated and semiautomated instruments have built-in programs for calibration (see operator's manual).
Biannually:	Clean or change all filters on washers. Check all fluid lines and tubing on support equipment or instruments for signs of deterioration and replace as needed; regrease and recalibrate all pipettes as needed (grease is usually provided with the pipette; see maintenance instructions).
Annually:	Change all fluid lines and tubing for major instruments; arrange for a service call by a factory representative if possible; recalibrate pipettes.

Pipettes are precise and important basic instruments of the laboratory, and as such, need to be calibrated at least annually. Several methods of pipette calibration are available, including the pipetteting of colored solutions followed by spectrophotometric analysis, the weighing of volumes of distilled water, or pipetteting volumes of radioisotopes. One commercial company (MLA, Pleasantville, NY) markets a spectrophotometric calibration kit for mechanical pipettes.

When selecting a pipette for a routine procedure, choose the pipette in the most appropriate range for the volume needed. Several ranges of volumes can be provided by two or three pipettes. Never exceed the range of the pipette as damage may result and precision may vary. Technicians should watch carefully while pipetteting to ensure that accurate amounts are delivered. It is not possible to detect minute differences in delivery volume simply by observation, however, occasionally a tip may not be well seated and a loose pipette tip may only pick up 50 vs. 100 µl, for example. Vigilantly observing the pipette tip as each specimen is added will prevent these errors. Multichannel pipettetors also warrant careful attention to assure that each tip is properly seated and that each barrel delivers the appropriate volume consistently and accurately.

For convenience, many laboratories make batches of wash buffers or stop solutions and store them in containers that dispense fixed volumes. When changing these solutions it is easy to accidentally disturb the volume setting, thus altering the delivery volume. To ensure accurate reagent delivery/consistent performance, recheck the volume dispensed after any disturbance. Also, these containers should be cleaned periodically before adding new reagent.

Most photometric instruments require calibration to ensure accuracy and linearity of their readings. This is usually accomplished using special calibration plates, available from the manufacturer. The plates consist of different wells, each capable of producing a different O.D. reading. Observed readings are compared to theoretical values and evaluated using confidence limits. Likewise, automated diluters must be calibrated. This is usually accomplished by diluting a standard color solution and reading O.D. spectrophotometrically. Results must be within 10% of the expected absorbance. The manufacturer of any automated instrument can be contacted for details of these procedures specific to their instruments.

VIII. COMMON ERRORS ENCOUNTERED DURING HIV TESTING

The following are a few of the more common errors that occur during HIV and retroviral testing:

1. Dilution errors. Carefully calculate and perform dilutions of all reconstituted components. Most ELISA kits contain instructions for dilutions, but many are based on volumes required per plate. When working with smaller numbers of strips or wells, or two or more plates, the required volumes are different, but the proportions remain the same (see Chapter 9 for a discussion of dilutions).
2. Scratching the coated antigen during sample addition. This is a common problem in both microtiter plate ELISA and IFA techniques. Do not touch the wells of the plate or slide with the pipette tips while adding samples or reagents.
3. Improper dropper use. In some agglutination assays, as well as some of the dot-blot tests, a dropper is supplied for adding reagents. Two important points should be remembered while using these droppers: first, they are to be used in a vertical position while adding reagents and second, free falling drops are essential. Failure to observe either of these practices may result in inaccurate volumes.
4. Using improper pipette tips. Be sure to use the proper tips for the pipettes, otherwise inaccurate volumes may result.
5. Inconsistent technique. Each specimen and control must be treated exactly the same. Do not treat controls more carefully than unknowns. Also, be sure that the same technician adds all controls and samples.
6. Use of improper equipment. Use the right equipment. If a high intensity lamp and magnification are required for reading an agglutination test, do not try to use the tests without them. Your results will not be reliable. Also, be sure that all equipment is properly calibrated.
7. Improper temperature. Allow kits and specimens to reach room temperature before use; adding a cold conjugate to a competitive binding assay would severely alter reaction kinetics, possibly producing false-positive results.
8. Mixing of components from different lots of kits. Reagents of each lot are titrated to work optimally as a kit. Many WB kits have conjugate dilutions that fluctuate from 1:500, 1:1000, 1:1500, etc., depending on

Quality Control and Quality Assurance

the given lot. In addition, the avidin-biotin conjugate systems may require different dilutions for each conjugate.

IX. TROUBLESHOOTING

A. General Assay Problems

When a test run fails, the reason may be readily apparent in some cases, but not so obvious in other instances. "Troubleshooting" refers to the measures used to determine why a run has failed.

The most effective way to prevent the failure of a run is to review the procedure carefully before beginning. Anticipate procedural steps ahead of time. For example, many conjugates require reconstitution and time to equilibrate before use. Be sure that one is not prevented from observing the proper incubation temperatures and times. Most importantly, follow the manufacturer's protocol precisely.

If an HIV test run fails after these precautions, an investigation must be conducted to determine the most probable cause. The following suggestions may help identify the cause of some common general problems:

1. Review protocol with the technician who ran the test, carefully checking for procedural omissions or changes.
2. Double check component expiration dates and the master lot expiration date.
3. Verify that all physical parameters of the assay were followed (times and temperatures).
4. Confirm that the support equipment such as pipettes, plate washing and reading systems, etc., are working properly. Verify that the preventive maintenance and servicing procedures have been performed.
5. Check the wavelength of the plate reader.
6. Investigate the calibration of the pipettes. If the assay calls for 10 µl of specimen, there is very little room for error in the volume delivered for the assay. If a pipette delivers 9 µl, this constitutes a 10% error and may be sufficient to cause the controls to fall outside acceptable limits. Proper calibration of pipettes is essential for accurate results.
7. Check the quality of the distilled water.
8. Check that reagents are not contaminated and were prepared and stored properly. There are several situations in which characteristic appearances in a test system may indicate problems. Examples include:
   - If the substrate tablets appear "orange" when they should appear "yellow to white" — this indicates that the tablets have been contaminated or have deteriorated, and thus cannot be used.
   - If the reagent "indicator cells" or "carrier particles" in an agglutination assay appear clumped or otherwise heterogeneous — this reagent may have been contaminated or may have deteriorated. Most package inserts describe the appearance of the reagents contained within the

kit, so ensure comparison of the actual appearance with what is expected.

If the intrarun reproducibility is not within the expected limits, the reason should be investigated. Instrument performance can be checked by performing electronic self-tests on ELISA washers, readers, and automatic diluters. Maintenance and recalibration of these instruments can be performed as often as required, although a factory service representative should be contacted to service any instruments requiring excessive recalibration and maintenance. Pipettes can be checked for accuracy and performance by weighing specific delivery volumes of distilled water (density = 1 g/ml). Pipettes are checked for precision by pipetteting a given volume of distilled water ten times and then weighing each volume on an analytic balance. The coefficient of variation between each weight obtained should be ≤15%.

B. Specific Assay Problems
Problem #1. All wells in an indirect ELISA are colorless upon completion of the procedure.

Possibilities: The conjugate or substrate may have deteriorated, or were added in the wrong order, or may not have been added. Note that the conjugate usually is the first reagent to deteriorate in an ELISA assay as the expiration date is approached. Conjugates can also deteriorate if the test kits become "heat spiked" during delivery, or are otherwise compromised due to improper storage conditions, etc. Also consider the conjugate dilution, a critical factor in assay performance. Substrates are usually stable if stored properly for the period of time recommended by the manufacturer; however, if the reconstitution is done improperly or the dilution is incorrect, color may not develop in the assay. Alternatively, a completely negative plate may reflect all negative samples and omission of the positive controls (or possibly, the samples may not have been added at all). These possibilities represent examples of the value of incorporating an external control into each run.

Solution: To prepare for situations such as this, it may be useful to retain a small aliquot of conjugate and substrate from each lot. When a failure of this type occurs, the aliquots of conjugate and substrate should be mixed to observe whether a rapid complete color change occurs. If a rapid color change does in fact occur, then the problem most likely is not due to either the conjugate or substrate and suggests technical error. Review procedure to ensure that all steps were followed. Try another kit from the same lot number; if this also fails, contact the manufacturer for a possible replacement kit.

Problem #2. All wells in an ELISA are the same color (in each row).

Possibilities: It is important to inspect the color of the substrate before adding it to the wells. If it is not colorless or is slightly yellow, or if it is otherwise not the color stated in the package insert, your entire run may be compromised.

Solution: Scrupulously clean glassware is highly important as residual material remaining in poorly washed glassware can react with the substrate. Similarly, if unbound conjugate is not completely washed out of the wells, it can react with the substrate, obscuring the "true" reaction of the substrate with bound conjugate. This is why it is important to ensure a good wash of the ELISA plate or wells following the manufacturer's protocol, and ensuring that no residual material remains to react. However, many manufacturers recommend that single-use disposable-type containers (such as polypropylene 15 or 50 ml centrifuge tubes) be used for preparing working dilutions of conjugate and substrate, thus avoiding contamination problems.

Problem #3. Upon visual inspection of the assay, there is a normal appearance to the reaction wells (i.e., the blank well is clear, positive and negative control wells are darker and lighter, respectively), yet when read on a spectrophotometer, all of the absorbances are nearly identical.

Possibilities: Either the detection or reference wavelength (or both) of the spectrophotometer are incorrect.

Solution: Correct the wavelength and then reread.

WB (WB) assays are also susceptible to technical problems (see WB Troubleshooting Guide, Appendix D). The following are a few of the most common problems in performing WBs.

Problem #4. WB strips fail to develop.

Possibilities: Conjugates were added in the wrong order, or the dilution was not correct. Substrate may have become nonfunctional. Positive controls may have been omitted.

Solution: If the substrate is suspect, rewash the strips and add fresh substrate. Otherwise, repeat the entire test.

Problem #5. Negative control strip has an unexpected band.

Possibilities: Contamination between wells or the reaction was allowed to develop too long.

Solution: Repeat the procedure with the same controls plus the negative control from another lot number. Put the controls in different places on the tray. If both negative controls have the same band, the problem is most likely contamination; if the negative control from the previous lot is still working properly, the new negative control should be discarded. If a contamination problem occurs in a WB system, and it cannot be attributed to carryover, all support equipment, including vacuum line hoses, automatic dispensing equipment, and any other item that comes in direct contact with the blotting system will either have to be thoroughly washed or replaced to eliminate any possible source of contamination.

C. Instruments

The most obvious indication of instrument error is when a control falls out of range. However, subtle changes in instrumentation such as with ELISA readers, washers, or pipettes may not be apparent until shifts or trends occur, requiring weeks to be detected.

Preventive maintenance is the best method for avoiding instrument failure. When there is an indication of failure, several measures can be undertaken.

ELISA Washers:

- Check all reservoirs and tubing to ensure an even flow of wash solution; and inspect lines for leakage.
- If there is a vacuum gauge, examine vacuum pressure to ensure that it is in the proper range.
- While the plate is washing, examine the ELISA plate at eye level to observe the tray for correct alignment and proper filling and aspiration of all wells. Some washers allow for user adjustment of the washer head alignment; however, most do not. In this situation it is necessary to contact service personnel. If one of the aspirating or dispensing jets is occluded, try gently probing the individual jet with a stylet. Be sure to rinse the washer thoroughly by priming and washing several times with distilled water. After this, run distilled water through the instrument for several wash cycles.

ELISA Readers:

- Many of the more recent models have electronic self-testing capability that can be performed by the user (see user's manual). This will only help diagnose the problem because electronic failures of the reader will require a service call from a factory representative.
- Double check the wavelength setting, the dual vs. single wavelength mode setting, and paper alignment.

D. Internal Control Problems

If the internal controls fall out-of-range, consult the run rejection rules in the package insert in the section describing positive and negative controls and validation of the assay. Sometimes one control value is permitted to be rejected without rejecting the run. In most cases a single out-of-range control can be attributed to technical error. If enough controls are out-of-range to cause run rejection, a more serious systematic error should be suspected and the cause investigated. The initial parallel check results for this lot can be examined to compare differences in performance. Another approach is to run the kit controls from the previous lot in parallel with the current lot, which would give an indication if there was a control contamination problem. Contamination could occur if a technician accidentally pipetteted negative control with the same tip used for the positive control. If contamination of the controls is expected, call the manufacturer and request new internal controls from the same lot; in most cases these can be provided.

If all controls are either high or low rejects, have a second technologist repeat the run. If the same results are obtained and the external control is also out-of-range, this indicates a system error rather than technical error, and again all possibilities regarding support equipment should be investigated. If all troubleshooting measures fail to identify the problem, the most likely source is a problem with the kit itself. At this point the manufacturer should be consulted. If there is truly a problem with the kit, the manufacturer will usually replace it.

E. External Control Problems
What action should be taken when an external borderline control falls within 2s of the mean but the actual O.D. is below the cutoff value?

Possibilities: There is a true systematic error, or the titer of the external control is too low.

Action: If the laboratory is still in the process of gathering data on the new external control, this is an indication that the titer of the control is too low. If the control continues to perform in this manner, the dilution is too great. Adjust the titer accordingly, and continue gathering data. The mean should be recalculated from the point of adjustment. Repeat the process until a reproducible value is obtained.

What if the external control falls "outside" of 2s?

Possibilities: As above.

Action: In this case, if control limits have already been established, and the control has performed adequately in the past, then the run must be rejected and repeated. If the same results are obtained on the repeated run, then technical error is most likely not the cause and the true cause must be investigated.

First, select a new vial of the same lot from the second freezer (if the lot vials were evenly divided between two freezers as suggested) and run both in parallel on the next run. This could indicate that the control has been contaminated or has deteriorated if one is acceptable and the other one is not. If so, discard the contaminated vial.

However, if both vials fail, this may indicate a problem with the entire lot. Double check freezer temperatures and storage conditions for any obvious signs that would indicate the cause of the failure. In a situation such as this, it is helpful to have a backup external control. Again, this control can be either purchased or prepared by the laboratory. If there is a suspected problem with the external control, run the backup external control with the current external control in parallel on several runs. Assuming the backup control performs adequately and the current external control still yields aberrant results, then there could be a contamination problem with the entire lot of control that somehow was not previously revealed. To confirm this suspicion, randomly select and test another vial of the same lot. If this vial also fails then the entire lot must be discarded and a new lot prepared.

It cannot be emphasized enough that the quality of the preparation for the external control must be scrupulously maintained throughout the preparation process to ensure that a homogeneous mixture is evenly distributed to all control vials. If the control performs well during the data gathering process, is stored properly, and is used as indicated, then the external control will perform well and serve the purpose for which it was intended: to monitor the overall performance of the assay and provide indications of QC failure that would otherwise be undetected.

REFERENCES

Blum, A. S., Computer evaluation of statistical procedures, and a new quality control statistical procedure, *Clin. Chem.*, 31, 206, 1985.

Dux, J. P., *Handbook of Quality Assurance for the Analytical Chemistry Laboratory*, Van Nostrand Reinhold, New York, 1986.

Damato, J., Fipps, D. R., Redfield, R. R., and Burke, D. S., The Department of the Army quality assurance program for human immunodeficiency virus antibody testing, *Lab. Med.*, 4, 577, 1988.

Epstein, J., Sensitivity and consistency of screening tests for antibodies to human immunodeficiency virus type-1, *Transfusion*, 31, 388, 1991.

Gerber, A. R., Valdiserri, R. O., Johnson, C. A., Schwartz, R. E., Hancock, J. S., and Hearn, T. L., Quality of laboratory performance in testing for human immunodeficiency virus type 1 antibody, variables associated in multivariate analysis, *Arch. Pathol. Lab. Med.*, 115, 1091, 1991.

Hutchison, D., Optimizing human-factor system reliability: an application of total quality management to AIDS testing, *Perspect. Lab. Manage.,* August, 1991.

Kudlac, J., Hanan, S., and McKee, G. L., Development of quality control procedures for the human immunodeficiency virus type 1 antibody enzyme-linked immunosorbent assay, *J. Clin. Microbiol.*, 27, 1303, 1989.

Lawsuits test liability for transfusion-related HIV infections, *Clin. Lab. Lett.*, 9, 20, 1991.

National Committee for Clinical Laboratory Standards, NCCLS Document C24-A, Villanova, PA, 11, 6, 1991.

Spiegel, M. R., *Statistics,* McGraw Hill, New York, 1961.

Westbrot, I. M., Statistics for the Clinical Laboratory, J. B. Lippincot, Philadelphia, 1985.

Westgard, J. O., Groth, T., Aronsson, T., Falk, H., and de Verdier, C. H., Performance characteristics of rules for internal quality control: probabilities for false rejection and error detection, *Clin. Chem.*, 23, 1857, 1977.

Chapter 9

LABORATORY TECHNIQUES

I. INTRODUCTION

The scope of QA extends beyond the laboratory testing process and includes the blood collection process, labeling, the proper selection and use of equipment, the methods for preparing dilutions, etc. Each of these is equally important in assuring the quality of final results. For example, it is extremely important to establish proper identification of the individual to be tested to ensure that the results are reported on the proper patient. Similarly, the specimen must be collected in the appropriate tube and all tubes must be fully labeled at the collection site. All information must be checked to ensure that the tube and laboratory request slip match. Clerical errors may cause results to be reported on the wrong person. Specimens must be transported to the laboratory under the proper conditions and in a timely manner. Incorrect preparations of reagents, or improperly diluted specimens may be just as detrimental to the quality of results as a test that is performed incorrectly.

II. SPECIMEN COLLECTION

A. Types of Specimens

Blood is collected from veins, arteries, or capillaries. The site of collection has no effect on the outcome of serological tests, although the peripheral veins of the antecubital space are often the site of choice (out of consideration for patient comfort). Phlebotomy is the term used to describe the opening of a vein for a blood collection.

Serum is obtained from blood after the blood has coagulated in a tube containing no anticoagulant. Serum does not contain fibrin, since fibrin is consumed in the coagulation process. It is separated from the red cell clot by centrifugation. A red top phlebotomy tube contains no anticoagulants and is the specimen of choice for HIV testing.

Plasma is obtained if blood is collected into a tube containing an anticoagulant. Most anticoagulants function by chelating calcium in the plasma; Ca^{2+} is necessary for several steps in the coagulation cascade. Table 13 lists the various types of phlebotomy tubes and the anticoagulants they contain. Because the coagulation cascade is not initiated, plasma samples contain fibrin.

Urine offers a unique alternative to serum and plasma samples for testing because it can be obtained noninvasively. Assays have been developed, but not yet FDA approved, that can detect HIV antibody in urine. However, HIV antibody concentrations are much lower than those found in blood, and there may also be a diurnal variation in the concentration of antibody. Therefore, the best time to collect urine samples for testing is the first morning discharge.

TABLE 13
Phlebotomy Tubes and Anticoagulants

Anti-coagulant	Stopper color	Chemical basis	Application
None	Red	Coagulation proceeds	Serology — ELISA,[a] WB[a]
EDTA	Purple	Binds Ca^{2+}	Hematology[a] Flow cytometry
Citrate	Blue	Binds Ca^{2+}	Coagulation[a]
Heparin	Green	Inhibits thrombin	Flow cytometry[a] Chemistry[a] Viral culture[a]
ACD	Yellow	Binds Ca^{2+}	Flow cytometry HLA typing

[a] Specimen of choice.

B. Blood Collection Methods

When collecting blood for testing, safety is the most important consideration, followed by proper procedure in order to collect a specimen that is adequate for testing. Practicing universal precautions includes wearing gloves and protective clothing while drawing blood (Chapter 11). When drawing blood from individuals for HIV testing, the most important consideration is to prevent infection from being passed from the individual to the laboratory worker, and from the laboratory worker to the individual. When drawing blood from many individuals, examination gloves should be removed, hands should be washed, and a new pair of gloves should be donned before handling the next patient.

When blood is drawn, hemolysis can be avoided by not leaving the tourniquet on the arm for more than 1 min. If it is difficult to find the vein and it takes longer than 1 min, the tourniquet should be removed and the circulation allowed to return 1 or 2 min before continuing.

1. Syringes

Blood is collected using a 21-gauge needle and either a 5-ml or larger syringe. For more difficult veins, the use of a "butterfly needle" (21 gauge × $^3/_4$ in. with 12 in. tubing, luer lock) will increase the amount of control and thus improve the chance for success. For those patients with fragile veins or veins that tend to collapse when vacutainer tubes are used, the syringe allows an extra measure of control over the speed at which blood is collected, thus reducing the possibility of collapsing veins.

Hemolysis can result when transferring blood forcefully through a needle into a vacuum tube, and therefore should be avoided. The proper procedure to avoid injury and hemolysis is to place the plastic cap of the needle on the bedside table or countertop, then using only the hand holding the syringe, guide the needle into the cap. Next, remove the cap and needle from the syringe. Finally, remove the rubber cap from the vacutainer and dispense the blood gently and directly into the tube. Replace the rubber cap on the tube for transport to the laboratory. Care should be taken not to exceed the tube's labeled capacity, otherwise the cap will not stay in place. In addition, excessive or inadequate amounts of blood will adversely affect the anticoagulant concentration, and may further interfere with testing.

2. Vacuum Tube Devices

The vacutainer type system used for drawing blood samples has some advantages over the use of syringes. When samples are drawn for many different laboratory tests, only one venipuncture is required to draw blood into several different tubes. Properly used vacuum tube systems do not cause hemolysis, because blood enters the tube at an even rate under vacuum. Vacuum tube systems may be more difficult to use when a patient with difficult veins is encountered.

3. Filter Paper As a Carrier for Blood Specimens

Collecting and processing blood specimens in remote laboratories or under field conditions is often troublesome. It may not be possible to separate the serum or to refrigerate the specimens during storage or transport. A technique has been developed for collecting whole blood that is particularly useful for large serosurveys and for pediatric specimens. A drop of whole blood obtained by a finger prick is absorbed onto a specially designed and standardized piece of filter paper (Schleicher and Schuell #903, Keene, NH, U.S.). The blood is allowed to dry thoroughly and can then be transported to the laboratory for immediate testing or stored for up to 3 months at room temperature without loss of antibody. The filter paper specimens can also be refrigerated or frozen, but must be sealed in plastic bags containing a desiccant.

For antibody testing, a prescribed circled area is punched out of the blood-spotted filter paper and placed in an elution buffer or diluent. A 6.3-mm disk completely saturated with dried blood contains 10 µl of blood or approximately 5 µl of serum. The disk is placed in a blank microtiter well on a clean microtiter plate (not on a test plate), and a specified diluent is added to elute the antibody from the filter paper, resulting in a dilution of serum. Eluates at proper dilution then can be used in modified ELISA and WB procedures.

Many of the commercial suppliers of ELISA and WB tests have adapted their protocols for use with whole blood specimens collected on filter paper. The manufacturer's recommendations should be followed precisely in order to assure that final dilutions of the specimen are correct. Only those tests that have been adapted and evaluated for filter paper methods should be used.

As with any laboratory method, careful quality control measures are critical. Control filter paper disks can be prepared with HIV-positive and -negative blood. These controls should be included in each test run. If possible, a set should be stored with specimens as a control for storage conditions.

III. SPECIMEN PROCESSING

A. Receipt of Specimens

Improperly labeled tubes should not be accepted by the laboratory. Furthermore, the laboratory staff should not attempt to correct labels on submitted specimens; this must be performed by the person who submits the specimen.

B. Acceptable Specimens

Quality results can only be obtained from specimens that are suitable for testing. A number of conditions interfere with the performance of laboratory tests, and thus make the results questionable. Questionable results are not to be reported; therefore, certain criteria for acceptability are necessary. Samples that are hemolyzed, lipemic, icteric, or contaminated with bacteria can yield false-negative or -positive results and must not be used. Some surgical procedures such as the "fluorescent angiogram" transiently contaminate the plasma with a fluorescent dye; a specimen such as this would be unsuitable for IFA testing because false-positives would occur. Therefore, it is very important to inspect all specimens prior to accepting them for laboratory testing. A notation stating the reason for rejection of the specimen, and a request for a new specimen should be returned to the submitter of any unacceptable samples.

C. Clotting and Centrifugation

Whole blood will clot within 20 to 60 min at room temperature, more slowly if kept in the refrigerator or on ice. The blood is then centrifuged at approximately 3000 rpm (approximately $1700 \times g$) for 15 min to separate serum from cells. Blood tubes should always be centrifuged with either the original stoppers left in place or covered with parafilm to prevent aerosols (see Chapter 11). Alternatively, blood can be left undisturbed at room temperature for a longer period of time, where separation by sedimentation will occur.

Antibody titer is not greatly affected in samples left at room temperature for up to 1 week; however, samples can become grossly contaminated with bacteria, even at refrigerator temperatures if nonsterile containers are used and sterile technique is not practiced. Therefore, serum specimens should:

- Not remain at room temperature for any longer than necessary (hours only)
- Be refrigerated as soon as possible
- Be frozen at $-20°C$, if not tested within 3 to 5 d

D. Pooling of Specimens for HIV Testing

The cost of HIV testing is prohibitive for many developing countries. One approach that has been suggested for reducing the costs of large-scale HIV screening is to pool aliquots of 3 to 10 serum specimens and test them as one. For surveillance, the prevalence can be directly calculated from the number of pools that test reactive; however, if individual test results are required, only the sera comprising the reactive pools need be retested individually. This technique is most cost effective for large-scale testing in which the prevalence of HIV is low (<5%). The technique of pooling should only be used with tests that require a diluted sample as the pooling itself results in a dilution. This method is not appropriate for use with competitive ELISA procedures.

An ELISA or agglutination test must be carefully modified for testing pooled sera, and thoroughly evaluated before being adopted by a laboratory. The following points should be considered:

- What is the largest number of samples that can be pooled without compromising accuracy due to specimen handling and recordkeeping?
- How many specimens can be pooled without sacrificing sensitivity?
- How many specimens can be pooled without decreasing specificity due to high background O.D. readings?
- Will a weakly reactive specimen be detected following dilution with other specimens that are negative?

Of course, this technique has its primary utility when conducting large serosurveys for epidemiological studies, and under no conditions can this be practiced for blood bank screening until further evaluations are performed and the method verified as acceptable.

IV. VOLUMETRIC MEASUREMENTS

The ability to accurately weigh and transfer precise volumes of sample and reagent is perhaps the most fundamental of all laboratory skills necessary for successful laboratory testing. Accuracy and precision in weighing, reconstituting, diluting, and pipetting of both reagents and samples during all steps of the testing procedure is critical for achieving valid results.

A. Pipettes

The pipette is the instrument used for volumetric measurements in all clinical diagnostic laboratories. Obviously, the manner in which the pipette is handled and maintained directly affects the accuracy of laboratory results. The three most common types of pipettes routinely used for HIV testing are transfer (or Pasteur pipettes), serological, and manual (or mechanical).

Glass Pasteur pipettes and plastic transfer pipettes are commonly used in many clinical laboratories for aspirating and transferring serum from clot tubes. While some may suggest that a greater volume of serum can be recovered using the glass Pasteur pipette, safety must take priority over increased serum recovery; therefore, plastic transfer pipettes should always be substituted for glass Pasteur pipettes in any laboratory testing situation because the danger of puncture wounds due to broken glass represents a great risk.

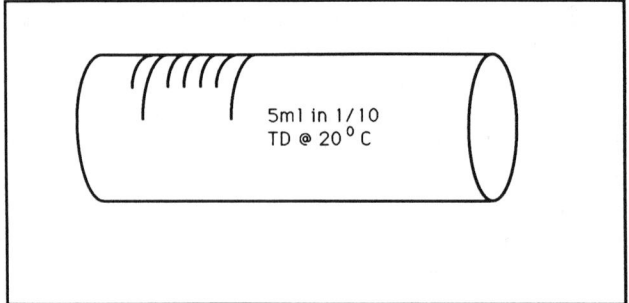

FIGURE 51. Pipette labels.

Serological pipettes are available in a variety of volumes, from 0.2 to 25 ml, and are routinely used when making dilutions or reconstituting reagents. Information about the pipette is usually printed at the aspirating end and indicates the total volume and units of delivery (Figure 51).

The pipette represented above will deliver a total volume of 5 ml. The information on the second line "TD @ 20°C" indicates: "To deliver at 20°C". (*Note: It is important to maintain an ambient room temperature of 20° to 25°C since extremes above or below this range will adversely affect the accuracy of the volumes delivered by pipettes, especially for enzyme-based tests with significant steps carried out at room temperature.*) Each large division demarcates 1 ml, while each fine division indicates 1/10 or 0.1 ml. The accuracy of a 5-ml pipette will be +/-0.1 ml.

Manual (mechanical) pipettes are the method of choice for pipetting small volumes. Single-channel (1 barrel) and multichannel (8 or 12 barrel) pipettes are available. Single-channel pipettes are available in ranges from 0.5 µl to 5 ml. Multichannel pipettes are available in ranges from 5 to 300 µl. Most of the mechanical pipettes operate on an air displacement principle. The plunger passes through two or three stops, depending on the manufacturer. From the release position to the first stop is the calibrated stroke and is determined by the volume setting (Figure 52). The area between the first and the second stop is used to blow out residual serum that remains at the pipette tip. Some pipettes have the ability to eject the pipette tip if the plunger is pushed to a third stop.

When using either single- or multichannel manual pipettes, it is important to avoid bubbles that alter volume. The following steps will help eliminate bubbles during pipetting:

1. Depress the plunger to the first stop (step 1, Figure 52), place the pipette tip just below the surface of the serum, then aspirate slowly (step 2, Figure 52). Slow aspiration is essential to obtain correct volumes. In addition, it will help prevent aspiration of liquid into the pipette barrel.
2. If the tip is inserted too far below the surface, a small serum droplet can adhere to the outside of the tip. Carryover of this excess serum will

Laboratory Techniques

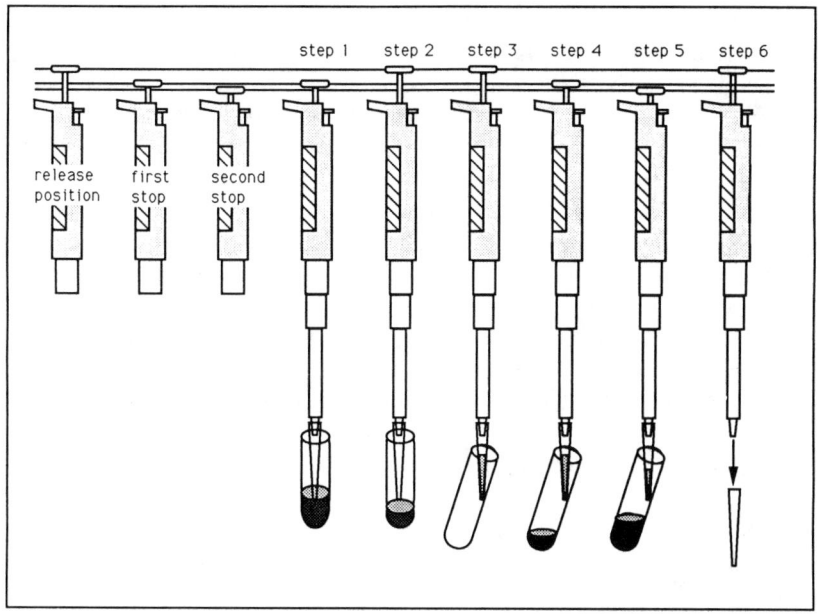

FIGURE 52. Single-channel manual pipette operation.

increase the actual volume delivered, which may alter the reagent concentration and ultimately the test result. This error in the volume delivery is increased as the desired delivery volume decreases.
3. Properly fitting tips will ensure trouble-free and accurate performance of a calibrated pipette.

1. Single-Channel Manual Pipettes

Important specimen delivery steps must be followed with any mechanical pipette. When dispensing a volume with bead assays or tubes, the tip should be placed against the side of the tube or well as close to the bottom as possible (step 3, Figure 52). The tip should not be placed directly on the bottom of the well in a vertical position, because this affects delivery due to blockage of the tip orifice. Delivering the volume at the top of the well or tube should be avoided, since the entire volume may not run down the side of the tube and will not make sufficient contact with the reactants. The tip should be placed against the side of the tube or well, very near the bottom, and the liquid allowed to flow to the bottom (step 4, Figure 52).

When dispensing a volume of sample, diluent, or conjugate into a well of a microtiter plate, it is important to get as close to the bottom of the well as possible without actually touching the bottom. Touching the well can scratch the antigen or antibody coat, which will alter the reaction. Serum should never be added directly to Ag- or Ab-coated wells unless specifically indicated by the manufacturer; diluent should be added prior to serum.

2. Multichannel Manual Pipettes

The addition of diluent or reagents to many consecutive wells with a single-channel pipette may introduce slight variations in volume, regardless of the level of technical skill. Consequently, uneven reaction conditions due to pipetting variability will occur, possibly affecting assay performance. Multichannel pipettes aid in overcoming these conditions by making repeated additions of the same volume of diluent or reagent to many wells an easier and more accurate procedure. As was the case with single-channel manual pipettes, when using a multichannel pipette to dispense a volume of diluent or conjugate into a well of a microtiter plate, it is again important to get as close to the bottom of the well as possible without actually touching the bottom. Touching the well may scratch the antigen or antibody coat, which alters the reaction. However, with multichannel pipettes, well-to-well cross-contamination can also occur.

B. Mechanical and Motorized Pipetting Aids

A variety of mechanical and motorized aids for serologic pipettes are available and will increase the control and safety of pipetting (Figure 53). Mouth pipetting is not acceptable, due to the great risk of acquiring infections by mouth, therefore these aids should be used.

C. Pipetting Technique
1. Serological Pipettes

It is common in many laboratories to find experienced technologists aspirating and delivering volumes with a serological pipette held at an angle. It is best to always aspirate and dispense reagents in the vertical position. The pipette should be viewed at eye level and the meniscus of the fluid observed in precisely the same manner from one volume to the next. Consistency in delivery technique is also equally important as this affects the precision of repetitive volumetric measurements. The expected accuracy and precision can be obtained if pipetting is performed carefully and consistently while using the correct pipette.

A pipette should be chosen in which the finest units of measurements are one significant figure beyond the volume to be measured when selecting a serological pipette. For example, the pipette in Figure 51 is perfect for placing 1 ml vol of diluent in test tubes for a serial dilution. However, this pipette is totally inappropriate for adding 200 µl (2/10 ml) of diluent to a dilution tube, since the accuracy is +/-1/10 ml (100 µl).

2. Manual Pipettes

Bubbles are often an annoying problem when using a multichannel pipette. Bubbles may be avoided by pipetting in the following manner:

1. Place sufficient volume in a reagent boat or trough to fill the entire plate.
2. Press the plunger to the first stop before aspirating the diluent or reagent.
3. Place the pipette tips below the surface of the reagent in the reagent boat.

Laboratory Techniques 177

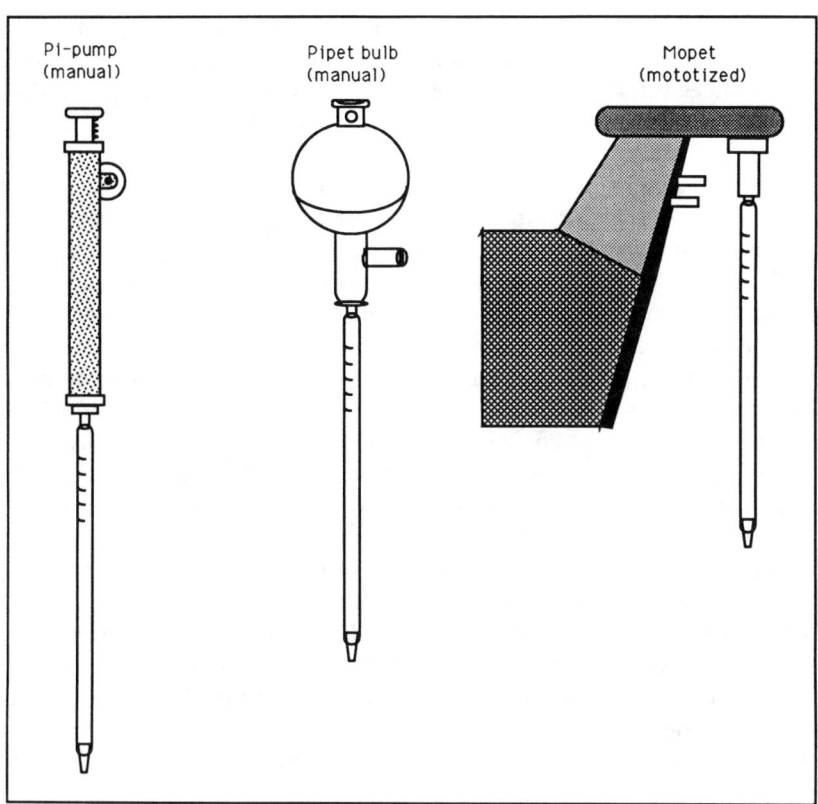

FIGURE 53. Examples of mechanical pipettes.

4. Raise the plunger slowly to the release position, keeping the pipette tip below the surface of the liquid.
5. Remove the pipette and dispense the liquid by slowly and evenly depressing the plunger to the first stop (do not release the plunger).
6. Finally, keep the plunger at the first stop while removing the tips from the wells. At this point, release the pipette plunger to the first stop, and use the pipette to aspirate additional reagent to continue the process.

Note: Most pipette manufacturers require that pipettes be depressed to the blow out position (if the pipette has such a position) before returning for more reagent. This increases the accuracy of delivering volumes, and should be performed if it can be done without producing bubbles.

If a pipette is used to transfer reagent to consecutive wells without changing tips, no residual reagent should remain in the tips. If this occurs, new tips should be placed on the pipette before addition of reagent to other wells. The most accurate method to ensure precision is to change pipette tips for each row.

Some debate exists as to whether a pipette tip should be "primed" or "wetted" before use. These terms refer to aspirating and dispensing the liquid several times before proceeding. Some consider that this "priming" practice increases volume precision. This may be valid for the multichannel pipette because a slight difference in volume will occur between the first and second row of the microtiter tray if the multichannel pipette tip is not wetted before adding reagent to the first row. However, for the single-channel pipette, the most important consideration when adding serum is consistency. Since the tip is changed for each specimen, it is much more important that all serum and controls be pipetted in a consistent manner. This point can be demonstrated when adding controls. Suppose an assay requires three negative and three positive controls. To save tips, the same tip is used for all three negatives, then using a single new tip the positive control is added to all three wells. If the tip was not wetted or primed before adding control to the first well of each series, well 1 will have slightly less volume than wells 2 and 3. This may produce a drop in O.D. sufficient to cause a low reject of the first control. To avoid this problem a separate pipette tip should always be used for each well, even when adding the same control to three different wells. Also, if a particular tip is damaged or manufactured incorrectly, the use of different tips will prevent all controls from being inaccurately delivered. By pipetting in this manner, consistency is insured from start to finish between controls and specimens. (Prewetting pipette tips when dispensing samples with a single-channel pipette is a tedious and time-consuming practice that is unnecessary. As mentioned above, it is much more important to consistently pipette in the same manner. A slight variation may occur in the volume delivered by a prewetted vs. an unwetted pipette tip, although this difference should not be significant to affect the final test result.)

D. Reagent Reconstitution and Glassware

When reconstituting reagents, note that glassware such as volumetric flasks will have a "TC" designation such as TC 250 ml @ 20°C. A volumetric flask with these designations will "contain" 250 ml at the prescribed temperature, yet will not "deliver" exactly 250 ml as the flask is emptied because residual amounts will adhere to the walls of the container. It is important to note designations on pipettes and glassware and select the appropriate vessel or serologic pipette. Some glassware such as beakers and Erlenmeyer flasks will not have a TC or TD designation because their intended uses do not require the precision indicated by this designation.

Most importantly, it is necessary to thoroughly read all information provided by the manufacturers to ensure that all pipettes and glassware are used in accordance with recommendations; periodically inspect, clean, and lubricate devices as recommended. Assay results cannot be expected to be accurate if the tools of the laboratory are not in good working condition, or if they are used incorrectly. Not only is it important to use proper technique when pipetting, it

Laboratory Techniques **179**

is even more important to pipette in a consistent manner with all samples, controls, and reagents. (Note: Glassware have been a fundamental part of laboratory equipment for more than a century. However, as a greater emphasis is placed on laboratory safety, many of these items are slowly being replaced by plasticware. Whenever possible, plasticware, disposable polypropylene containers, and plastic disposable serological pipettes should be utilized to avoid the possibility of injury due to broken glass.)

V. PREPARING DILUTIONS

A. Simple Dilutions

Basic laboratory investigation requires that laboratorians be well educated in preparing solutions and preparing samples prior to the actual testing. Obviously, if any test is to perform optimally and yield accurate results, the reagents and samples must be prepared correctly. Although most commercial kits for HIV testing contain reagents that require very little preparation, some reconstitution or mixing is usually necessary. In addition serum samples are almost always diluted prior to testing. Usually the package insert will indicate the correct volumes of reconstitution fluid or diluents to add; these steps must be performed carefully and accurately.

With certain serologic assays samples must be diluted prior to addition to the test system. A notable example is the IFA test which usually requires a predilution of 1:20 in phosphate buffered saline (PBS). The aim of the examples below is to make the concept of dilutions more understandable by showing how dilutions can be manipulated to save time and reagents and how to eliminate unnecessary steps.

When a serum sample is collected from a patient, many times the HIV test is not the only laboratory test requested; the sample must be shared with other laboratory departments for additional testing. With a small sample volume, the conservation of as much serum as possible for repeat and confirmatory HIV testing is critical. When performing HIV or any other serological assays, never use less volume than is required for an assay, as this will produce invalid results. When a sample with insufficient volume is submitted to the laboratory for testing, notify the submitter and indicate on the request form "Quantity Not Sufficient" (QNS) for testing. However, if dilutions are required, the general principle is to use the minimum practical volume of serum possible when making all dilutions.

To prepare a dilution, a volume of the specimen is mixed with a certain volume of a buffer (or diluent). In a simple case, one part serum and one equal part diluent will yield a 1:2 dilution. The expression "1:2" represents the ratio of the number of parts of serum to the total number of parts, or one part serum and one part diluent for two total parts. Likewise a 1:100 dilution can be prepared by mixing 1 part serum with 99 parts diluent; for example 1 ml + 99 ml (100 total parts), or 50 µl and 4950 µl (4.95 ml) can also be used.

B. Serial Dilution Technique

When making a 1:100 dilution, 1 ml is an excessive volume of serum to be diluted. A 50-μl volume is a more conservative amount of serum to use, but the volume of diluent required may not be practical with a heavy workload. In these and similar situations serial dilutions are the practical choice for preparing dilutions.

Serial dilutions involve mixing and transferring a constant volume of serum into successive tubes containing diluent. When working with dilutions it is helpful to express all dilutions in terms of fractions. This makes solving a serial dilution problem a matter of multiplying fractions. For example, a 1:2 dilution may also be expressed as the fraction:

$$\frac{1}{2} = \frac{\text{Parts of serum}}{\text{Total parts (serum + diluent)}}$$

A twofold serial dilution is a dilution where a fixed volume of serum is mixed and transferred into successive tubes containing the identical volume of diluent. This dilution fold can be expressed as:

$$\frac{1}{\text{Dilution fold}} = \frac{\text{Parts transferred}}{\text{Total parts}}$$

Example #1 — A Twofold Dilution (Figure 54)

Place 100 μl of diluent into 8 tubes. Add 100 μl of serum to the first tube. Mix the contents by gently moving the pipette plunger up and down 3 to 4 times, then transfer 100 μl of this volume to the next tube. Repeat the process until the final tube is reached. The last volume can either be discarded or left in the tube. The pipette tip does not need to be changed after each transfer step (for routine serologic testing). However, when serially diluting standards or sera for quantitative assays, or when preparing dilutions for viral isolation, many authorities recommend that pipette tips should be changed after each dilution step to prevent carryover (carryover could falsely elevate the titer). For each tube above:

$$\frac{\text{Parts transferred}}{\text{Total parts}} = \frac{100 \ \mu l}{200 \ \mu l} = \frac{1}{\text{Dilution fold}} = \frac{1}{2}$$

This represents a twofold dilution.

The twofold dilution shown above yielded a final dilution of 1:256 and involved the use of 8 dilution tubes. Using a constant 100 μl volume to mix and transfer, the final dilution of 1:256 can also be made by several different manipulations of diluent volumes.

Laboratory Techniques

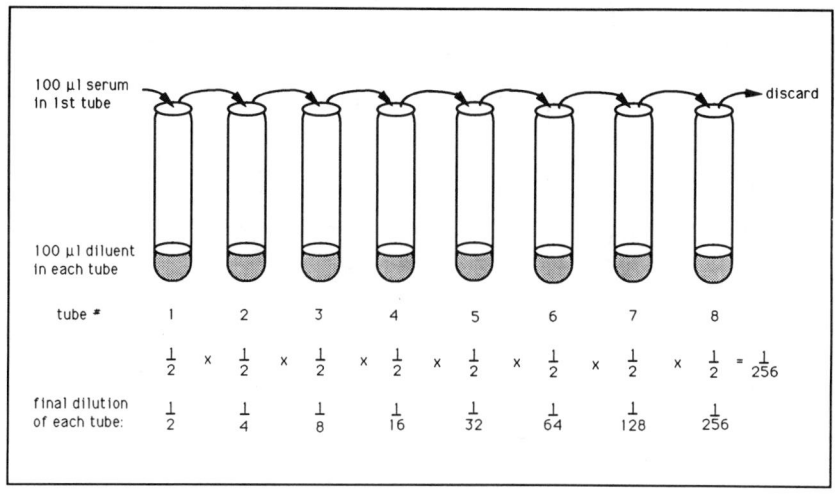

FIGURE 54. Dilution tubes (Example 1).

Example #2 — A Fourfold Dilution (Figure 55)

Set up 4 tubes with 300 µl of diluent in each. Add 100 µl of serum, mix and transfer 100 µl as directed above. The total volume of 400 µl in each tube makes a 1:4 dilution in each tube.

Example #3 — A 16-Fold Dilution

This dilution can be made by placing 1500 µl (1.5 ml) of diluent in 2 dilution tubes. Add, then mix and transfer 100 µl of serum as before.

$$\frac{1}{16} \times \frac{1}{16} = \frac{1}{256}$$

Example #4 — Varying the Diluent Volume

In this example, it is necessary to quantitate a positive HIV IFA result. On the previous day, 3+ fluorescence was observed at a 1:20 dilution. What would be the most appropriate range of dilutions to use to find the endpoint at 1+ fluorescence? What would be the most efficient way to make these dilutions?

Ordinarily, when positive samples are quantitated by fluorescence the brightness of the apple-green fluorescence is graded on a scale of 1+ to 4+, with 4+ being the strongest. The titer can be estimated by knowing that for each twofold increase in dilution the degree of fluorescence will diminish by approximately one unit of brightness (observation by the authors). In this case the estimation is that the endpoint dilution would be approximately a 1:80 dilution. The most efficient way to approach this situation is to ensure that the least amount of wells on the HIV IFA slides are used, and to select the most appropriate range

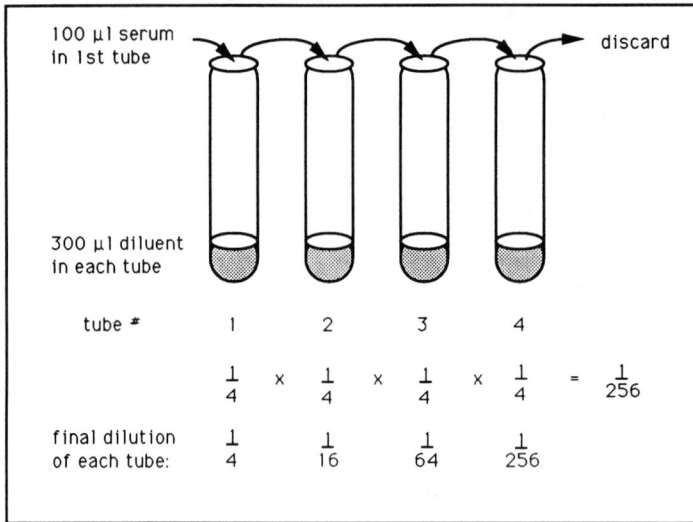

FIGURE 55. Dilution tubes (Example 2).

(as to avoid the necessity of repeating the test). Therefore, a 1:40, a 1:80, and a 1:160 dilution of the positive sample should be tested. However, these are general guidelines and further dilutions may be necessary before an endpoint reaction can be obtained. One dilution alternative would be to mix and transfer 100 µl of serum into tubes as shown in Figure 56.

In this alternative, six tubes were necessary to obtain the desired dilutions. It should be noted that any slight error in the dilution process is exponentially increased with each additional tube. Therefore, it is always best to use the minimum number of tubes to obtain the desired dilutions. A more efficient way of obtaining the same dilutions and using only four dilution tubes is represented in Figure 57.

In summary, when making serial dilutions, one must recall that the volume of serum used to mix and transfer remains constant, while the diluent volume is easily changed to yield the most appropriate dilution. Also, to conserve serum, always use the smallest possible volume to dilute. However, remember that the use of <50 µl (final volume in each dilution tube) is not recommended for serial dilutions, because a volume less than this would be difficult to control accurately. (Author's suggestion: Many errors occur when making dilutions because researchers lose track of the dilution sequence. This can be avoided by repositioning each tube in the rack following the addition of fluid. With microtiter plates, a plate cover can be used, or a marking can be made on the bench and the plate can be moved across the mark on the bench after each addition to indicate the next well to be used.)

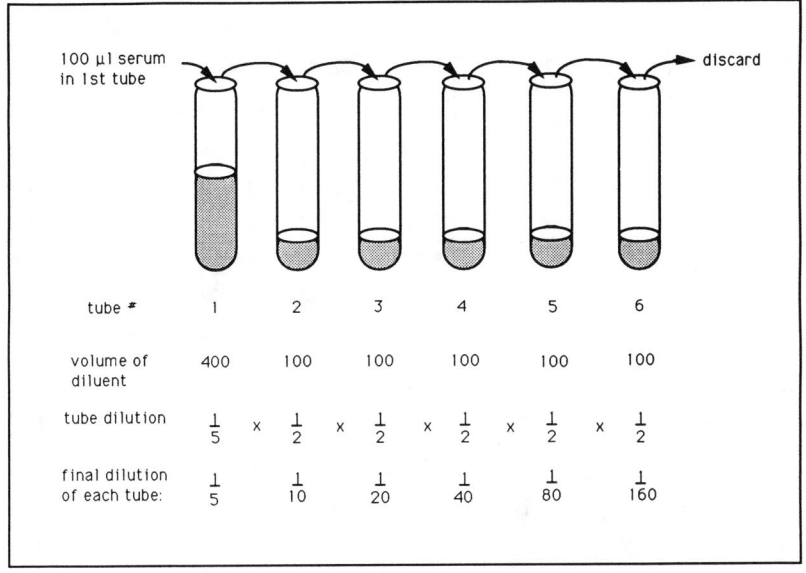

FIGURE 56. Dilution tubes (Example 4a).

VI. THE TESTING OF PREPARATIONS INTENDED FOR HUMAN INJECTION

Several countries have policies in effect that require the testing of any imported blood product intended for human injection for the presence of the HIV antibody. These products often include blood products such as anti-Rh immunoglobulin, factor VIII, and other pooled blood fractions. Although the intent of such a policy is commendable, several reasons exist as to why it is ineffective in accomplishing the intended goal of certifying that a given preparation is incapable of transmitting HIV infection. The result, however, may be a false sense of security or a waste of valuable products. Listed below are several points that should be considered when testing preparations other than normal human sera:

1. Serologic tests for anti-HIV were designed to perform optimally on fresh normal human sera. The validity of results on products other than sera are disputable because test performance may be altered due to the different concentrations of proteins.
2. The absence of antibody to HIV does not guarantee that a product is not infectious. Antibody may not always be detectable even when the virus is present. Likewise, the presence of antibody does not indicate that a sample is capable of transmitting HIV. This can only be confirmed by virus isolation.

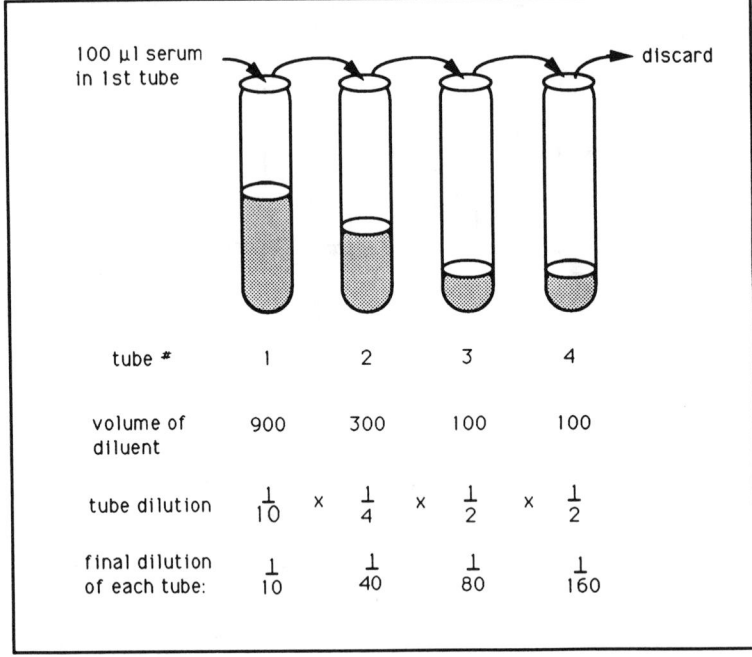

FIGURE 57. Dilution tubes (Example 4b).

3. If the product is purchased from a reputable manufacturer, the process of preparation of these injectables virtually eliminates any chance of survival of viable virus. All units used to contribute to the pool of "anti-Rh immunoglobulin" (for example) are individually screened prior to inclusion in the preparation. All seropositives are eliminated.
4. Approximately 500,000 doses of Rh_o-Ig are given annually to 350,000 women in the U.S., and no evidence exists to implicate anti-Rh immunoglobulin as a source of HIV infection.
5. Pooled immunoglobulin and other preparations usually do not contain the same concentration of immunoglobulin as is found in NHS. Therefore, these preparations should not be evaluated using conventional screening methods.
6. If these preparations are tested, the limitations of such testing should be reported together with a disclaimer stating that test results may not be valid.

A. Dilution of Concentrated Products

Immunoglobulins or other protein preparations (such as factor VIII) for human injection are purified, lyophilized, and usually concentrated. To optimize the chances of detecting HIV antibody, the concentrated preparation should be reconstituted as directed by the package insert and then diluted with specimen diluent. The dilution should approximate the normal human serum concentration of either immunoglobulin or total protein (depending on the type of preparation). To determine the amount of dilution that should be made of reconstituted preparation, first convert the given concentration to milligrams per deciliter. The concentration of immunoglobulin in NHS is approximately 900 to 2100 mg/dl, while the concentration of total protein is approximately 7 g/dl, depending upon the population. For the purposes of this discussion, assume 1500 mg/dl for immunoglobulin, and 700 mg/dl for total protein, respectively.

Example #1
Problem: A Rh_o-Ig product is labeled 300 µg/ml; find the concentration in mg/dl (1 dl = 100 ml). (Note: 1 µg = 1×10^{-6} g.)

$$\frac{300 \text{ µg}}{\text{ml}} \times \frac{1 \text{ mg}}{1000 \text{ µg}} \times \frac{100 \text{ µl}}{1 \text{ dl}} = \frac{300 \times 100}{1000} = \frac{300}{10} = 30 \text{ mg/dl}$$

30 mg/dl is much more dilute than 1500 mg/dl (immunoglobulin), and therefore the product is tested without making a prior dilution. Results are reported with a disclaimer citing limitations as above.

Example #2
Problem: A factor VIII product is labeled 700 mg/ml; find the concentration in mg/dl

$$\frac{700 \text{ mg}}{1 \text{ ml}} \times \frac{100 \text{ ml}}{1 \text{ dl}} = 70,000 \text{ mg/dl } (70 \text{ g/dl})$$

This product will require a dilution before it approximates the concentration of total protein in NHS. (Recall that 700 mg/dl is used.)

$$\frac{70,000 \text{ mg/dl}}{700 \text{ mg/dl}} = 100$$

Therefore, this product should be diluted 1:100 before testing.

If the laboratory should happen to receive an enzyme preparation for HIV testing, note that enzymes are measured in units of activity (IU or mIU) rather than concentration. These are not easily converted to milligrams per deciliter.

Remember that results of these preparations must be reported with a disclaimer mentioning test limitations on products other than NHS. The caveat often used in these circumstances is "Specimens such as pooled human sera or enzyme preparations are not recommended for testing by commercial assays for HIV-antibody. Therefore, interpretation of the results is difficult, and results may not be valid. Also, since these assays detect only antibodies to HIV, a conclusion cannot be made concerning the presence of infective virus."

REFERENCES

Diagnostic tests for HIV, in *AIDS 89 Summary, A Practical Synopsis of the 5th Int. Conf. on AIDS*, Vol. 4, Philadelphia Sciences Group, Philadelphia, 1989, 115.

Farzadegan, H., Quinn, T., and Frank, B., Detecting antibodies to human immunodeficiency virus in dried blood on filter papers, *J. Infect. Dis.*, 155, 1073, 1987.

George, R. J., Hannon, H. W., Jones, W. K., Hoff, R. et al., Serologic Assays for Human Immunodeficiency Virus Antibody in Dried Blood Specimens Collected on Filter Paper from Neonates, U.S. Department of Health and Human Services, Centers for Disease Control, Atlanta, 1989.

Henry, J. B., *Clinical Diagnosis and Management by Laboratory Methods*, W. B. Saunders, Philadelphia, 1991.

Kaplan, L. A. and Pesce, A. J., *Clinical Chemistry: Theory, Analysis, and Correlation*, C.V. Mosby, St. Louis, 1984.

Mitchell, S. and Mboup, S., HIV testing, in *The Handbook for AIDS Prevention in Africa*, Family Health International, 1990, 20.

Centers for Disease Control, Lack of transmission of human immunodeficiency virus through Rh(D) immune globulin, *MMWR*, 36, 728, 1987.

Chapter 10

HOW TO EVALUATE DIAGNOSTIC TEST KITS

I. INTRODUCTION

Chapter 8 mentioned that part of a TQM system included the use of the best available tests. Therefore, laboratories may need to evaluate new tests periodically as they become available, or change tests as need dictates.

A large number and variety of retroviral diagnostic assays are commercially available, many of which offer characteristics that are attractive as well as essential. New and improved tests are always being developed, as noted in the extensive list of commercially available retroviral tests (Appendix A). Reports in the literature and package inserts from the commercial companies suggest that these tests have extremely high performance levels. Chapters 3 through 6 discussed the advantages and disadvantages of some of these tests, and indicated the different reasons that a particular test may be more useful in certain diagnostic situations. Chapter 7 included a discussion of the various parameters that are often used to determine how accurate, reliable, and useful a test is, and how the test indicators are calculated. It was stressed in the earlier chapters that tests do indeed perform differently in various testing situations (conditions in the laboratory, technical expertise, populations being tested, etc.). Therefore, it is imperative that when evaluations of test kits are to be conducted, they be performed under the same conditions, and in the same location in which the test will ultimately be performed.

Certain basic guidelines must be followed to ensure that tests are evaluated properly. This will prevent incorrect information about a test from being reported, and will result in a fair and unbiased opinion concerning a manufacturer's product. It may also result in the communication of important information to the manufacturers, so that they have an opportunity to subsequently modify their products in an effort to improve the ability of the tests to achieve optimal performance in all testing situations. The following discussion of guidelines for the proper evaluation of retroviral tests is short, but important enough to warrant a separate chapter.

II. REFERENCE TESTS

Obviously, if any test is to be compared to another test, a means must exist to determine which test actually produces the most accurate results. Simply because a test has been used by many laboratories, or used for a long time, does not ensure that it produces the most correct results, or provides an efficient testing method for the laboratory. An apparently false-positive result by a new test may actually be a true positive result because the routine test may be producing a false-negative result. In order to determine which test is most appropriate to use, an established reference test must be used as a basis for comparison. If two tests are being compared, there must always be a "tie breaker"; however, even in this case, a third test may also produce an incorrect result. Therefore, the *number* of different tests used do not necessarily indicate the true status of the samples being tested (i.e., the quality of results).

A reference test is one that has been accepted by the medical community as a highly accurate and dependable test. It must possess a reputation as being highly accurate, and its results must be sufficiently correlated with the actual status of the patient. Only then can other tests be compared and evaluated. Currently, the reference test in serologic testing for retroviral infections is considered to be the WB. However, even this test may yield inconclusive results, and additional tests may be required to elicit the true status of the sample. These tests include synthetic peptide based tests, PCR, and viral culture. Therefore, the WB usually can be considered the reference test, but occasionally must be used in combination with more sophisticated technologies, if available.

III. CHOICE OF SPECIMENS

To evaluate serologic tests, panels of well-characterized specimens must be used. Well-characterized sera are those in which the true immune status of the individual is known with a high degree of certainty; i.e., the individual is infected or not infected based on clinical data or detailed and conclusive laboratory investigation. If conclusive information is not available, then the samples must be characterized using as many serological tests as possible. Alternatively, samples should be sent to reference laboratories for characterization. The following guidelines indicate the choice of specimens for use during an evaluation of HIV tests:

1. Panels of sera should include reactive, nonreactive, weakly reactive, and indeterminate-classified sera. If HIV-2 infections occur in the area (or are suspected to occur) in which the test will be used, several HIV-2 reactive sera should also be included in the panel.
2. At least 50 specimens should be included in each panel. Commonly, many more are used, especially negative samples. If sufficient numbers of sera are not available, as many as possible should be used. Note that the greater the number of sera used in each panel, the more statistically valid is the comparison.
3. Specimens used for the evaluation should be derived from individuals in the same geographic location, and from a representative sample of the same populations where new test will be used.
4. If specimens are not freshly collected, they should be stored frozen in aliquots. The use of aliquots is necessary to ensure that samples stored

under equivalent conditions are tested by each method; i.e, it is not appropriate to test samples by one of the methods after an additional freeze-thaw cycle.
5. All specimens must be in good condition for testing, i.e., nonhemolyzed, nonlipemic, no bacterial growth, etc.

IV. TESTING CONDITIONS

Each test being evaluated must be performed under identical conditions. For example, the performance of all tests must be conducted: within the same laboratory, using the same equipment (pipettes, instruments, etc., if the assay permits), on the same day (if possible), by the same technician, using kits that have not expired and have been stored properly, and using the exact protocols established by the manufacturers. In addition, the individual performing the tests must not know the results that were obtained from any of the other tests; i.e., the specimens tested by each assay are tested blindly.

V. DISCREPANT RESULTS BETWEEN ASSAYS

If discrepant results between the tests occur, they should be investigated further. Steps in this process include:

- All specimens yielding discrepant results must be repeated using all tests that are being evaluated. This will ensure that technical error did not occur. Repeat testing should be performed under the conditions stated above.
- If specimens still exhibit divergent results, those specimens should be retested by both tests, and by a different technician. The second technician must not know the previous results (i.e., the specimens are tested blindly). If possible, an aliquot of the sample should be sent to another reference laboratory and tested by the same assays blindly. This will ensure that the results were not technician or laboratory dependent.
- If the tests being compared do not include a reference test, all discrepancies must be tested using such a test, and not compared solely in relation to a third test that has not been established as a reference test.
- When a discrepancy occurs between tests and the sample cannot be conclusively characterized by the reference test (e.g., indeterminate by WB), the results from that sample must not be used in the evaluation since its true status is not known. For example, if test A produces a reactive result, test B produces a nonreactive result, and the WB indicates p24 reactivity only, the true status of the sample is unknown and the results of that sample should not be used in the calculations regarding a conclusion in the evaluation. This exemplifies the need to use well-characterized samples when performing an evaluation.

In conclusion, the evaluation of tests requires a systematic approach and careful interpretation of results. Well-characterized and appropriate samples must be used, and testing must be performed by well-trained individuals with a knowledge of the method(s) by which an evaluation should be performed. The testing conditions must reflect those under which the test will be ultimately performed, and reference tests must be used.

REFERENCES

Constantine, N. T., Fox, E., Abatte, E. A., and Woody, J. N., Diagnostic usefulness of five screening assays for human immunodeficiency virus in an East African city where prevalence of infection is low, *AIDS*, 31, 313, 1989.

Crofts, N., Maskill, W., and Gust, I. D., Evaluation of enzyme-linked immunosorbent assays: a method of data analysis, *J. Virol. Methods*, 22, 51, 1988.

Griner, P. F., Mayewski, R. J., Musklin, A. I., and Greenland, P., Selection and interpretation of diagnostic tests and procedures, *Ann. Intern. Med.*, 94, 557, 1981.

Maskill, W. J., Crofts, N., Waldman, E., Healey, D. S., Howard, T. S., Silvester, C., and Gust, I. D., An evaluation of competitive and second generation ELISA screening tests for antibody to HIV, *J. Virol. Methods*, 22, 61, 1988.

Report of the WHO Meeting on Criteria for the Evaluation and Standardization of Diagnostic Tests for the Detection of HIV Antibody, Stockholm, Sweden, WHO/GPA/BMR/88.1, World Health Organization, Geneva, 1987.

Chapter 11

SAFETY IN THE LABORATORY

I. INTRODUCTION

Throughout the history of laboratory practice, the issue of biosafety received relatively little attention, although published reports of laboratory-acquired illness appeared as early as the turn of the (20th) century. It was not until the recognition of AIDS, and the subsequent testing for HIV that began in 1985, that safety considerations became widely addressed. Awareness of the laboratory risks, as well as knowledge of the means needed to prevent infection in a laboratory setting, have dictated the need for safety as an important part of the laboratory function.

Common sense safety practices are the principal means by which laboratory personnel can avoid infection from biohazards. Brown and Blackwell may have expressed the current biosafety situation most accurately by saying, "Gloves and skill are all that stand between laboratorians and the viral, bacterial, and infectious agents they handle all day." A safe working environment in the laboratory can be maintained by instituting biosafety practices that also include careful measures for personal hygiene, for cleanliness in the work space, and for proper handling of biohazardous materials. An understanding of these necessities and an awareness of potential biohazards while performing laboratory duties will help to prevent accidents, injury, and infection.

II. OCCUPATIONAL RISKS

A. Infectious Material

The CDC has recommended that gloves be worn when blood, CSF, amniotic and pericardial fluids are handled. However, in a reassessment of the original position, their recommendation for universal precautions has been dropped for a second category of human material that includes feces, semen, saliva, tears, breast milk, vomitus, urine, and cervical secretions, unless there are obvious signs of blood.

The risk of exposure to HIV, although significant, is not as acute as the risk of exposure to hepatitis B virus (HBV). HBV is transmitted in the same manner as HIV, is found in many of the same body fluids as HIV, and is the most commonly occurring form of laboratory-acquired infection. Neither virus has been demonstrated to be capable of causing infection through aerosols. It has been reported that some categories of laboratory workers have seven times the risk of exposure to HBV as does the general population. For these reasons, it is recommended that all health care workers receive HBV vaccines. Infection by either HIV or HBV is a significant threat in the laboratory setting, and preventive means to protect against either will also control and protect against the other.

B. Accidental Exposure

The primary work-related dangers are parenteral exposure through accidental needle sticks, cuts from contaminated equipment, exposure of mucous membranes to aerosolized droplets, and exposure of chapped or broken skin, wounds, and scratches to contaminated specimens.

Although many potential routes of exposure exist, it has been reported that up to 80% of all exposures occur as the result of needle sticks. In any case of exposure of a health care worker, the source of exposure (patient) should be identified, and the patient informed and tested for antibodies to HIV after consent is obtained. If the patient should refuse to be tested, the CDC recommends that the worker seek counseling and evaluation by a clinician (some states permit testing without consent in such situations). Seronegative health care workers who have been exposed should be followed for 6 months, with testing at 6 weeks, 12 weeks, 6 months, and finally at 1 year postexposure. If seroconversion has not occurred by 1 year after exposure, then the worker was most likely not infected.

In 1983, the CDC began a Cooperative Needlestick Surveillance Group to follow up accidental exposures. After a participant is documented as negative for antibodies to HIV, they are enrolled anonymously, assigned a number within their institution, and then monitored for seroconversion. Employee health representatives may enroll an exposed employee by contacting the CDC at (404) 639-1644. This program is also applicable to firemen, policemen, or other emergency service personnel who have been occupationally exposed and meet the criteria for enrollment.

C. Work Space

"Containment" of infectious agents is the principal means of providing a safe laboratory working environment. Containment can be accomplished through the adoption of a combination of policies and safety regulations that govern everyday work practices. Some authorities recommend that work space should be divided into "contaminated" vs. "clean" areas to effectively utilize protective equipment such as gloves, lab coats, or safety eyewear. Labels should be attached to instruments, or signs may be posted in areas indicating risk due to contamination. If the laboratory has more than one sink, one should be designated the "clean" sink for handwashing. Liquid biohazardous waste such as ELISA or WB wash waste should never be disposed of in the handwashing sink.

Lab coats and gloves must be used in the contaminated area when handling any equipment, pipettes, specimens, refrigerators, and freezers where samples are stored. Protective equipment must also be used when working under a hood (class II biosafety cabinet). Gloves and laboratory coats should be removed, and hands washed before handling any clean equipment such as the computer, calculator, or telephone. A policy that limits access to laboratory testing areas

Safety in the Laboratory 193

can be instituted by indicating that "authorized staff only" be admitted; this will significantly limit unauthorized pedestrian traffic, thus avoiding unnecessary contact of individuals with contaminated equipment.

A clean, organized work space is a safer work space. The following measures will help increase safety in the laboratory:

1. Reagents should be well labeled, in proper containers, and stored properly (example: store acids and bases in a separate cabinet below benchtop level).
2. Keep benchtop free from clutter (nonessential materials).
3. Materials should not be placed near the edges of counters or shelves, but rather organized as related to need.
4. Hazardous materials and reagents not in use should be safely stored out of the immediate work area.

Countertops or work areas should be cleaned with disinfectant before and after each work day using one of the disinfectants listed in Table 14.

Other prudent measures and safety practices that help reduce the risk of infection include:

- Immediate containment of spills and wiping of the contaminated surface using one of the disinfectants listed in Table 14.
- Avoiding the use of sharp instruments whenever possible (scalpels, needles, scissors).
- The use of pest control measures that will prevent known and unknown vectors from carrying disease in or out of the laboratory.
- The use of unbreakable materials.
- The cleaning of doors, handles, telephones, etc., with disinfectant.
- The availability of a first aid kit in every laboratory.
- The reporting of accidents immediately to the supervisor.
- The presence of an eye wash station for accidental exposure from splashes.

D. Biohazardous Waste

Proper handling of infectious material prevents infection. The following practices can assist in preventing exposure to infectious agents:

- Biohazard warning signs must be plainly posted, and should always be used.
- All laboratory waste must be properly decontaminated before disposal. This can be accomplished by autoclaving at 121°C at 1 atm for 20 min, or by incineration.
- Infectious waste should not be buried.
- A separate puncture-resistant container for glass waste and needles must be used.

TABLE 14
Recommended Disinfectants

Disinfectant	Final conc.
Sodium hypochlorite	0.1–0.5% available chlorine
(household bleach)	10%
Ethanol	70%
Isopropyl alcohol	70%
Formaldehyde	4%
Glutaraldehyde	2%
Hydrogen peroxide	6%
Povidone iodine	2.5%

- Spills must be immediately decontaminated with one of the disinfectants listed in Table 14.
- Liquid wastes should be decontaminated by autoclaving or by adding sufficient sodium hypochlorite before discarding.

III. UNIVERSAL PRECAUTIONS

Practicing "universal precautions" is the most effective and efficient method of preventing exposure to HIV and other pathogens. Universal precautions are applied in the care of *all* individuals, since medical history and examination cannot reliably identify those individuals who are infected with HIV. All blood and body fluids should be handled as if infectious, and all practical precautions should be observed. The decision to wear gloves when handling specimens that are not included in the list recommended for use of universal precautions should be made by the individual. Good hygienic practice is sufficient reason to wear gloves when handling samples.

The proper method of self-decontamination after performing work is to remove the lab coat first, then gloves, and finally to thoroughly wash hands with soap or disinfectant in the "clean" sink. If hands are washed before removing the coat, the hands may become contaminated again while handling the coat.

Recently, the U.S. Occupational Safety and Health Administration (OSHA) has become involved in the issue of laboratory safety by mandating that all laboratory employers must:

1. Educate and train all health care workers in the practice of universal precautions to prevent blood-borne disease
2. Provide the equipment necessary for such protection (i.e., gloves)
3. Monitor compliance with mandated biosafety practices

To provide a safe working environment in the laboratory, a comprehensive and continually upgraded policy regarding the containment of infectious ma-

Safety in the Laboratory 195

terial must be systematically enforced to ensure maximum safety to all workers. Observing the following hygienic practices while working in the laboratory will help to minimize exposure to pathogens associated with blood and body fluids.

The following practices should always be employed:

- Universal precautions should be practiced with all specimens at all times (i.e., treat all specimens as if they were infectious).
- Lab coats should always be worn while in the laboratory to ensure that clothes are kept free from infectious agents.
- Lab coats should not be removed from the laboratory area, except for laundering.
- Lab coats should be kept away from dining or food areas.
- Cuts and scratches on hands and arms should be covered and protected.
- Hands should always be washed after removing gloves, and before leaving the laboratory.
- Eating, drinking, and smoking in the work areas is a serious health risk and must not be practiced.
- Long hair should be pinned up or covered.
- Eyes should always be protected when a chance exists of generating droplets or aerosols.
- Activities such as applying makeup and combing hair should not be practiced in the laboratory.
- Protective shoes that completely cover feet should be worn (no open-toed shoes or sandals).
- Gloves should be worn when handling blood or bodily fluids.

REFERENCES

Barman, M.R., AIDS precautions in practice, *Med. Lab. Observer*, April 1990.
Brown, J. W., Biosafety in the laboratory, Testrends, Roche Diagnostic Systems, Inc., Company publication, Vol. 5, 1, 1991.
Brown, J. W. and Haider, M., The risk of AIDS to laboratorians: an update, *Med. Lab. Observer*, April 1990.
Brown, J. W. and Blackwell, H., Putting on gloves in the fight against AIDS, *Med. Lab. Observer*, November 1990.
Centers for Disease Control, Agent summary statement for human immunodeficiency virus and report on laboratory-acquired infection with human immunodeficiency virus, *MMWR,* 37, 1, 1988.
Biosafety in Microbiological and Medical Laboratories, U.S. Department of Health and Human Services, Centers for Disease Control, Atlanta, National Institute of Health, Bethesda, MD, 1988, 2.

Mitchell S. and Mboup S., HIV testing, in *The Handbook for AIDS Prevention in Africa,* Family Health International, Durham, NC, 1990, 20.

Weiss, S. H., Goedert, J. J., Garner, S. et al, Risk of human immuno-deficiency virus (HIV-1) infection among laboratory workers, *Science,* 239, 60, 1988.

World Health Organization, Report of a WHO Consultation on the Prevention of Human Immunodeficiency Virus and Hepatitis B Virus Transmission in the Health Care Setting, WHO/GPA/DIR/91.5, WHO, Geneva, 1991.

World Health Organization, Guidelines on Sterilization and High-Level Disinfection Methods Effective against Human Immuno-Deficiency virus (HIV), WHO AIDS Ser., WHO, Geneva, 1988.

SUMMARY AND CLOSING REMARKS

HIV and retroviral testing present a major challenge to the laboratory worker. Specimens must be handled, tested, and results must be interpreted and reported without error. One false-positive or -negative result due to negligence or technical error is unacceptable and can have dramatic consequences for the individual being tested or the recipient of blood. In addition to the normal care and expertise required to perform laboratory tests, the correct diagnosis of HIV infection requires a systematic and thorough QA program.

Technology is constantly evolving, and laboratory tests for the retroviruses are increasing in number and variety. As this technology progresses and competition increases, tests improve, and more alternatives become available. Although the tests that were initially introduced (viral lysate-based ELISA and WBs) are still widely used today, simpler and more rapid assays, as well as more sophisticated tests, have been added to the list. As shown in Appendix A, over 100 tests from 40 commercial manufacturers allow a wide range of tests for almost all possible testing situations. The continued need for improving tests is dictated by several existing problems, including early diagnosis, the resolution of indeterminate results, better confirmatory tests, less expensive tests, tests that are simpler to perform and are more foolproof, tests that can detect variants of the viruses, tests that include built-in QC measures, and tests that can detect infection in newborns.

Newer technologies introduce a means to address some of the problems still associated with the diagnosis of retroviral infections. Antigen/antibody complex dissociation agents attempt to increase the sensitivity of antigen tests when used to test serum. Class-specific antibody assays and substrate amplification methods may provide the ability to detect low levels of antibody in specimens such as urine and saliva. The third generation antigen sandwich ELISAs allow for the detection of all classes of antibodies simultaneously, thereby helping to detect early infection. Recombinant and synthetic peptide-based tests offer increased sensitivity and specificity, and can be used in the LIAs, or they can be applied to augment viral lysate WBs. Combination assays address the increasing problem of multiple retroviral agents and provide a means to test for more than one of these agents simultaneously and most effectively.

Assays are available to test for antibody, different isotypes of antibody, neutralizing antibodies, viral antigen, viral enzymes, viral DNA or RNA, and live virus. In addition, tests can be used to monitor retroviral disease progression, and include those for cell populations and subpopulations, unrelated products such as β_2 microglobulin and neopterin, and cytokines. Tests can detect antibodies to HIV-1, HIV-2, HTLV-I, HTLV-II, and combinations of all of these agents simultaneously; even HIV variants can be detected. Antigens used in the antibody tests have been extended to include recombinant proteins, synthetic peptides, and combinations of recombinant and synthetic peptides. Tests have become available that can be used for testing urine, saliva, whole blood, and capillary blood collected on filter paper. Total assay time ranges from over 24 h, to <2 min. Test formats include microtiter plates, beads, microbeads, dipsticks, dots, slots, paper strips, plastic strips, combs, swabs, cartridges, and combinations of these. Indicator systems include the production of color, an inhibition of color, radioactivity, fluorescence, dyes, and chemiluminescence. At least five different types of ELISAs are available, including indirect, competitive, antibody sandwich, antigen sandwich, and class-specific antibody capture types. Agglutination assays incorporate latex, gelatin, RBCs, and microparticles. Techniques include manual, semiautomated, and completely automated systems. Finally, the cost of each test can vary from $0.25 to >$150 (U.S.).

In this book, basic information is provided about the retroviruses, the tests currently used to diagnose infection, the interpretation of results, and the necessary QA measures needed to ensure that test results are obtained and reported as accurately and correctly as possible. With a thorough educational background in these principles, the laboratorian will have the skills necessary to produce the best quality results using the most appropriate technology. This will be the laboratorian's contribution toward the prevention and control of retroviral infections, particularly HIV infection and AIDS.

GLOSSARY

Activated lymphocytes: Lymphocytes that have been stimulated by contact with a pathogen or a cytokine.

Activated macrophages: The stimulation of mature macrophages, especially in response to phagocytosis or lymphokine activity.

Agglutination: The clumping of cells or particles due to lattice formation, as occurs when antigen-coated particles complex with antibody.

Agglutination assay: An immunoassay in which antibody reacts specifically with antigen-coated particles or cells to form a lattice network, resulting in the clumping or agglutination of the particles.

Algorithm: A sequence in which different diagnostic tests are used to arrive at a presumptive or definitive diagnosis.

Amino acids: The chemical building blocks of proteins.

Antibody: Proteins produced by the immune system in response to the presence of a foreign substance, an antigen or immunogen, with the capability to react specifically with the inducing antigen or immunogen.

Antibody titer: The concentration of antibody expressed as the reciprocal of the dilution. (For example, when a sample can be diluted 1:160 and still be reactive, but further dilutions result in a nonreactive result, the titer is 160.)

Antigen: Foreign substance that may induce the immune system to produce antibodies. These antibodies can specifically react with the antigen.

Antigen capture assay: An immunoassay to detect antibody in serum. A monoclonal antibody is usually used to capture the viral antigen and attach it to the solid support. The remaining part of the assay can be a competitive or an indirect ELISA for antibody detection.

Antigen detection ELISA: An enzyme immunoassay that has specific HIV antibody bound to the solid support. HIV antigen present in the sample, usually serum or tissue culture fluid, is bound to the antibody and detected by a conjugate-substrate system.

Antigenemia: The presence of viral components or antigens in the blood.

Antigen presentation: A process occurring by antigen presenting cells, whereby antigenic proteins are presented to T-cells in a form that they can recognize. For example, the gp120 envelope of HIV-1 is modified and presented on the macrophage cell surface along with the major histocompatibility complex class II molecule for recognition and interaction with the appropriate T cell receptor.

APC: Antigen presenting cell. A cell that phagocytizes a pathogen and processes the antigenic proteins for presentation to T-cells; they are usually macrophages.

ARC: AIDS-related complex. A stage of HIV infection in which the patient exhibits some signs of disease but does not meet the criteria for a diagnosis of AIDS.

Augmented Western blot: A WB test that uses a combination of viral lysate antigens and recombinant and/or synthetic peptide antigens.

Autologous red blood cell test: A test in which RBCs from the individual being tested are used as the solid support.

Autoradiography: A technique in which X-ray film is exposed to a radioisotopically labeled substance.

Avidin-biotin conjugate: An enzyme indicator system in which two separate reactants are used in order to increase sensitivity for demonstrating antibody in a specimen.

β_2-microglobulin: A structural part of the major histocompatibility complex molecule found on many cell surfaces. Elevated serum levels are noted as infection with HIV progresses.

B cell (B lymphocyte): The precursors of antibody producing plasma cells.

Biologic false-positive or -negative: A false test result due to parameters related to the specimen or patient, not to the testing process.

Borderline control: An external control that reacts just above (indirect ELISA) or just below (competitive ELISA) the cutoff point. A borderline control is used to detect minor changes in the test that may affect the accuracy of test results.

Borderline reactor: A test result that is just above or just below the cutoff point. The borderline reactor specimen should be retested.

Glossary

Capture ELISAs: ELISAs in which either antibody or antigen is trapped in order to increase the assay sensitivity.

CD4: A receptor molecule present on helper T lymphocytes and other cells that binds HIV. It is also a marker that identifies certain cells such as T_h cells.

CD8: A marker present on T lymphocytes that identifies them as suppressor or cytotoxic cells.

Chemiluminescence: An indicator system in which flashes of light are produced by a substrate and can be detected by a luminometer or on photographic film.

Chemotaxis: The movement of cells in response to a chemical signal. It is the process whereby phagocytic cells are either attracted or repelled by substances exhibiting chemical properties to the vicinity of invading pathogens.

Class-specific antibody capture assays: Tests that trap immunoglobulins (usually IgG) before detection, in order to increase the sensitivity of detection.

Coefficient of variation: The degree to which numerical data tend to disperse about the mean; an indicator of reproducibility of values.

Colony stimulating factor: A group of cytokines that affects the differentiation and maturation of progenitor cells into mature cell phenotypes.

Combination assays: Tests that have the capability to detect antibodies to more than one virus simultaneously.

Competitive ELISA: An enzyme immunoassay in which the unknown antibody in the sample competes with the antibody-conjugate for the same reactive sites on the antigen. The resulting color change is inversely proportional to the amount of antibody in the sample.

Confidence interval: A statistical estimate of the precision of some parameter.

Confirmatory testing: Tests performed on samples that are reactive by screening tests to ensure that the screening test results were accurate. Also called supplemental testing.

Conjugate: As used in the ELISA, an enzyme coupled or linked to an antibody. As used in the IFA test, a fluorescent dye linked to an antibody.

Conserved region: Sequences within the viral genome that do not vary, or vary minimally, from viral isolate to isolate.

Core: The internal region of the virus, bound by a protein coat, containing the RNA genome and viral enzymes, such as RT, integrase, and protease.

Criteria: Guidelines for determining if a test result fulfills the requirements to be labeled as positive, negative, or indeterminate.

Cross-reaction: The reaction of an antibody with an antigen other than the one from which the immune response was stimulated.

Cutoff (C.O.): A reference value for determining whether the O.D. reading obtained by an ELISA test represents a reactive or nonreactive result. The cutoff point is usually calculated based on the mean of the negative control values.

Cytokine: A generic term for nonantibody proteins, such as lymphokines or monokines, that are released by certain cell populations in response to specific antigens or cell injury. Cytokines act as intercellular mediators in the generation of the cell-mediated immune response.

Cytokine loop: A feedback mechanism in which cytokines react to cause an increase in their own production.

Cytotoxic cells: CD8+ T lymphocytes that recognize specific antigens on infected or malignant cells, and kill those cells.

Delta values: Calculated values that describe the ability of a test to resolve reactive and nonreactive sera.

Dilution: Mixing of one part of a substance with one or more parts of another substance, thus reducing the concentration of the first substance.

Dot-blot assays: Immunoassays that use paper or nitrocellulose as the solid support, and incorporate a dot of viral antigen blotted on the support.

EIA: Enzyme immunoassay. An assay that uses enzymes and substrates as the detection or indicator system. ELISAs are a type of EIA, but the terms ELISA and EIA are usually used synonymously.

Electrophoresis: A technique for separating proteins, usually based on their different electrical charges, and/or molecular weights (SDS-PAGE).

Glossary

ELISA: Enzyme-linked immunosorbent assay. An assay in which antigen or antibody is bound to a solid support and unknown antibody or antigen in the specimen is then specifically bound and detected by an enzyme-substrate system.

***env*:** Refers to the gene that directs the production of the envelope glycoproteins, such as gp160, gp120, and gp41 of HIV-1.

Envelope: The outer shell of a virus, composed of glycoproteins and lipids.

Enzyme: An organic catalyst. In the ELISA test, the enzyme acts upon a specific substrate to create a colored end product that is measured, and reflects the amount of enzyme bound in the reaction.

Epitope: The simplest form, or the least number of amino acid determinants present on a larger, more complex antigenic molecule that will combine with a specific antibody or T cell receptor.

External controls: Control sera, with known values or reactivity, derived from a source other than the test kit. These controls are included in test runs to monitor performance; lot-to-lot variation, etc.

Flow cytometry: A method to evaluate cell populations and subpopulations using an instrument known as a flow cytometer, which detects cells based on size, shape, granularity, and by the detection of cell markers.

Fluorescence: Emission of a longer wavelength of radiation by a substance that results as a consequence of absorption of energy from a shorter wavelength of radiation, such as UV light.

Fluorochrome: A substance such as fluorescein isothiocyanate (FITC) or phycoerythrin (PE), which fluoresces when exposed to UV light.

***gag*:** The group-associated antigen gene that encodes the core proteins of retroviruses.

Gene products: Proteins (antigens) encoded by the genome that are the structural and regulatory components of a virus.

Genome: The sum of all the genes of a virus.

Glycoproteins: Proteins coupled to carbohydrates or sugars.

Gray zone: The test results falling within a defined zone immediately below the C.O. value. Results falling within this zone often warrant further testing.

Helper T lymphocytes: A type of T cell that stimulates immune responses. Helper T lymphocytes cooperate with B cells and other T cells in initiating antibody-mediated responses. Also referred to as T4 or CD4+ T cells because they contain the CD4 marker.

HIV-1: Human immunodeficiency virus type 1, associated with AIDS. Initially called HTLV-III or LAV-1.

HIV-2: Human immunodeficiency virus type 2, also associated with AIDS and sharing partial homology with HIV-1.

HTLV-I: Human T lymphotropic virus type I, associated with ATLL, HTLV-I-associated myelopathy, and TSP.

HTLV-II: Human T lymphotropic virus type II, has been linked with hairy cell leukemia, but disease association has not been well established.

IFA: Indirect fluorescent antibody assay. An assay that uses a fluorochrome-labeled antibody to indirectly detect antibody or antigen.

IFN: Interferons. A class of low molecular weight proteins that exhibit antiviral activity. IFN is classified into three groups (alpha, beta, and gamma), depending on their cellular origin.

Immune complexes: Antigen-antibody complexes.

Immunocompetence: A term to describe the immune system when it is functioning properly.

Immunogenic: That which generates an immune response, usually an antigen.

Indeterminate: A result that cannot be classified as positive or negative based on the test results; also called an inconclusive or nondiagnostic result.

Indirect ELISA: An enzyme immunoassay for detection of antibody in which the antigen is attached to the solid phase, specific antibody in the serum is allowed to bind to the antigen, and an anti-human conjugate is allowed to react with the antibody. A substrate is subsequently modified by the enzyme, resulting in a colored product in direct proportion to the amount of specific antibody in the serum.

Glossary

Inducer cell: CD4+ T lymphocytes that induce other immune cells to mature, proliferate, or respond.

Infection: The stage at which virus has attached, penetrated, and been incorporated into the host cell.

Inflammatory cells: Cells that are recruited by chemotaxis in response to an antigen or injury. Cells included in this group are neutrophils, T-lymphocytes, monocytes/macrophages.

In-house controls: Control sera of defined reactivity and which are prepared in the investigator's own laboratory. In-house (external) controls are often prepared by pooling several specimens together, determining the test result, and aliquoting for routine use as a control.

***In situ* hybridization:** A technique in which radio- or enzyme-labeled HIV RNA is used to probe mononuclear cells for the presence of HIV DNA.

Interleukins: A group of lymphokine and polypeptide protein mediators secreted by leukocytes that facilitate cell-to-cell intercommunication.

Internal control: The positive and negative control sera included in a test kit.

Interrun reproducibility: A measure of repeatability between test runs.

Intrarun reproducibility: A measure of repeatability within one test run.

LIA: Line immunoassay. An assay, similar to the WB, which utilizes a plastic strip impregnated with antigens, usually recombinant or synthetic peptide.

LTR: Long terminal repeat. Gene sequences at the ends of the genome that may function to help regulate the expression of other viral genes.

Lymphocyte: A mononucleated WBC. There are several populations and subpopulations of lymphocytes.

Lymphocyte phenotyping: Test procedure to characterize lymphocytes based on their physical characteristics (e.g., cell surface markers).

Lymphokines: Soluble protein products that are released by lymphocytes upon activation by certain antigens or other cytokines.

Macrophages: Phagocytic tissue cells derived from the peripheral blood monocytes that have an accessory role in cell-mediated immunity.

Major histocompatibility complex: A group of genes that express cell surface proteins enabling cells to distinguish "self" from "nonself".

Mean: In HIV testing, the average value of a set of O.D. values or O.D./C.O. values.

Mitogen: A substance, usually a plant lectin, which has the ability to cause mitogenesis or induce proliferation, maturation, and activation of cells.

Monoclonal antibodies: Identical copies of a "single clone" of antibodies produced after immune stimulation that contain one H chain class and one L chain type, and are usually reactive with one epitope.

Monokines: Soluble protein products that are released by monocytes and macrophages upon stimulation by certain antigens or other cytokines.

Negative: The reported result of a confirmatory or supplemental test, indicating that the test did not detect evidence of viral infection.

Negative predictive value: A statement of the probability that a negative test result truly represents noninfection.

Neopterin: A soluble substance released from macrophages that is associated with macrophage activation, and may indicate disease progression.

Neutralizing antibodies: Antibodies that have the ability to combine with viruses and inactivate them, or help to prevent them from infecting other cells.

NK cells: Natural killer cells. Cells that are nonspecific killers of certain tumor and virus-infected cells.

Nonreactive: The reported result of a screening test, indicating that the test did not detect evidence of viral infection.

Nonspecific antibodies: Antibodies that may be bound in an immunoassay, but are not directed specifically to the viral antigen. In tests for retroviruses, nonspecific antibodies usually represent antibodies to nonviral or related viral proteins.

O.D.: Optical density. The units expressed by an ELISA reader representing the light adsorbed by the colored end product of the reaction.

O.D./C.O. ratio: The test optical density reading divided by the cutoff value for that run.

Glossary

Opportunistic infections: Infections occurring in an immunocompromised host that would normally not occur in a host with a competent immune system.

PCR: Polymerase chain reaction. A technique in which specific regions of the viral genome are amplified.

Plasma cells: A fully differentiated, activated B lymphocyte that produces antibody specific for a particular antigen.

PMN: Polymorphonuclear leukocytes. WBCs that are granulocytic and phagocytic.

pol: The polymerase gene or polymerase proteins.

Positive: The reported result of a confirmatory or supplemental test indicating that an individual has been infected by the virus.

Positive predictive value: A statement of probability that a positive test result truly represents infection.

Precursor molecule: A molecule produced early in infection and later cleaved to produce other proteins.

Predictive values: Description of the value of tests taking into account the actual prevalence of infection in the population being tested.

Primer: A specific sequence of amino acids that can be annealed to a matching region on the HIV genome for extension during the PCR assay.

Profile: The pattern of bands exhibited on a WB test strip, representing the specific antibodies present in the serum of an individual.

Provirus: The viral genome that has been integrated into host cell DNA, and which subsequently directs the synthesis of specific viral components.

Prozone: Inhibition of agglutination and resulting in a falsely negative test result at very high antibody concentrations.

QA: Quality assurance. The overall program ensuring that the final results reported by the laboratory are as accurate as possible.

QC: Quality control. Those measures that must be included during each assay to verify that the test is working properly.

Quality assessment: A means to determine the quality of results from a laboratory. It is usually an external system to check laboratory performance.

Parallel testing: A comparison of performance of new lots of kits with the previous lots via a common control material. Parallel testing is also performed on serial samples to directly compare results on the same run.

Passive agglutination: An agglutination assay in which the antigens are bound passively to the carrier, without specific attachment procedures.

Proficiency panel: A set of well-characterized sera sent to laboratories for testing in a routine manner. The laboratory test results are compared to the known values to determine the quality of the laboratory's results.

PHA: Passive hemagglutination assay. An agglutination assay that uses red blood cells as the carrier system.

Random error: Error due to chance, not necessarily that of the assay or the technician.

Reactive: The reported test result from a screening test that indicates presumptive evidence of infection.

Receptor molecule: The molecule on the host cell to which the virus specifically attaches.

Recombinant antigen: Antigens produced when a portion of the viral genome is inserted and replicated in a biological vehicle, resulting in the production of the gene product.

Reference test: A test that is accepted by the scientific community as a standard or reliable test.

Regulatory genes: Genes that are responsible for the production of components of a virus that modify viral expression and replication.

Retroviruses: Viruses that use an enzyme to transcribe an RNA genome into DNA.

RIPA: Radioimmunoprecipitation. A technique that utilizes a radioactive label to identify antibody to structural components, particularly envelope antigens.

Glossary

RT: Reverse transcriptase. The enzyme used by retroviruses to transcribe RNA into DNA.

Run: A group of specimens tested together in one batch.

Sandwich ELISAs: ELISAs based on the principle of antibody detection by trapping it between two antigens (antigen sandwich) or detecting antigen by trapping between two antibody molecules (antibody sandwich).

Sensitivity: The ability of a test to detect very small amounts of analyte, or the ability of a test to detect infected individuals.

Serial dilutions: Mixing and transferring a constant volume of serum into successive tubes containing diluent.

Seroconversion: The point at which specific antibody can first be detected in an infected individual who was previously negative for antibody.

Seroprogression: Evolution of the antibody response; an increase in antibody titer or antibody type over time.

Seroreversion: The point at which antibody can no longer be detected in an infected individual.

Shift: Rapid changes in the values of control sera, as noted on a QC graph. Shifts may indicate that a major change has occurred in some parameter of the assay.

Significant error: Error that is not random, but occurred due to a technical error, or due to some parameter associated with the test.

SOP: Standard operating procedure. A book containing detailed protocols for all procedures performed within the laboratory.

Specificity: The ability of the test to identify all negatives correctly.

Spectrophotometer: The instrument used to measure absorbance of light at specific wavelengths. An ELISA reader is a spectrophotometer.

Standard deviation: A measure of variation defining the expected range of a value in relation to the mean value.

Structural genes: Genes that are responsible for directing the production of components that determine the physical characteristics of a virus.

Substrate: A specific chemical that is modified by an enzyme, usually resulting in the formation of a colored product. Also used to describe the fixed cell layer in the IFA test.

Supplemental testing: Tests performed after screening tests to confirm the results of the screening assays (see Confirmatory testing).

Suppressor cells: CD8+ T lymphocytes that inhibit, suppress, or regulate the immune response.

Syncytia: Fusion of cells, usually immune cells, as a result of viral infection.

Synthetic peptide: Peptides created in the laboratory by combining amino acids in a known sequence to form the desired peptide, usually an antigen.

T4 lymphocyte (CD4+): A helper/inducer lymphocyte carrying the CD4 receptor molecule; also called T4 cells.

T cells: Cells that are produced in the bone marrow and migrate to the thymus where they multiply and mature into lymphocytes that are capable of an immune response. They play a significant role in cell-mediated immunity and are composed of several subpopulations.

Test efficiency: The overall ability of a test to correctly identify positives and negatives. The combination of the sensitivity and the specificity of an assay gives an idea of the total effectiveness of the assay.

TNF: Tumor necrosis factor. A cytokine, most often from macrophages, that may be associated with the pathogenesis of retroviral infections.

TQM: Total quality management. An approach to supplement quality assurance measures in which the efficiency and general operation of the laboratory are addressed and continually improved.

Transcriptional measures: Those measures that deal with the transfer of results from an instrument to worksheet, or from a worksheet to the computer or final report form.

Transmembrane: gp41 of HIV-1 or gp36 of HIV-2; the glycoprotein antigen located between the inner and outer membranes of the envelope.

Glossary **211**

Trend: Subtle changes in test values in the same direction, as detected by a QC graph; usually indicates that a gradual change has occurred.

Troubleshooting: Identifying problems in the test system and solving them.

Universal precautions: Precautions taken by health care workers to avoid being infected in the health care setting. Since medical history and examination cannot reliably identify all of those who are infected with an agent, all blood and body fluids are handled as if they were infectious.

Variable region: Sequence within the viral genome that may vary between viral strains.

Vigilance: Watchfulness in the laboratory, in order to identify potential problems.

Viral antigen: Components of a virus. Usually refers to the detection of antigens in serum or culture fluid (such as p24 in HIV-1 infection).

Viral lysate: Disrupted virions, often used as the antigen in viral immunoassays.

WB: Western blot. An enzyme immunoassay performed on antigens that have been previously separated by electrophoresis. Detects antibody to specific viral proteins and is considered as a confirmatory or supplemental test for antibodies to the retroviruses.

Window period: The period after infection has occurred but before antibody can be detected.

APPENDIX A
Retroviral Tests[a]

Name of test	Manufacturer (alphabetical)	Type of Test	Antigen
HIV ELISAs			
For the Detection of Antibody to HIV-1			
Abbott HIV AB[b]	Abbott Laboratories	I	L
Abbott Recombinant HIV EIA	Abbott Laboratories	I	RP
Enzygnost Anti-HIV micro	Behringwerke	C	L
Recombigen HIV-1 EIA[b]	Cambridge Biotech	I	RP
Karpas Cell Test HIV I & II	Cambridge Virucells	I	IC
Retro-Tek HIV ELISA[b]	Cellular Products	I	L
DuPont HIV-1 Recombinant	DuPont de Nemours	I	RP
DuPont ELISA[b]	DuPont de Nemours	I	L
ELAVIA I	Diagnostics Pasteur/Sanofi	I	L
HIV-1 ELISA	Diagnostic Biotechnology	I	L
LAV EIA[b]	Genetic Systems/Sanofi	I	L
REC VIH-KCO1	Heber Biotec	I	RP
Select HIV	IAF Biochemical/Coulter	I	SP
HIV-1 env Peptide EIA	Labsystems	I	SP
Ortho HIV ELISA System[b]	Ortho Diagnostic Systems	I	L
Vironostika Anti HIV-1[b]	Organon Teknika	I	L
Vironostika Uni-Form	Organon Teknika	C	L
HIV-TEK G	Sorin Biomedice	I	L
MicroTrak HIV-1 EIA	Syva	I	RP
UBI ELISA HIV-1 EIA[b]	United Biomedical International	I	SP
Wellcozyme HIV Recombinant	Wellcome Diagnostics	C	RP
For the Detection of Antibody to HIV-2			
HIV-2 AB	Clonatec	I	SP
ELAVIA II	Diagnostics Pasteur/Sanofi	I	L
HIV-2 ELISA	Diagnostic Biotechnology	I	L
Select - HIV	IAF Biochemical/Coulter	I	SP
HIV-2 EIA[b]	Genetic Systems/Sanofi	I	L
For the Detection of Antibody to HIV-1 and HIV-2			
IMx	Abbott Laboratories	F	RP
Recombinant HIV-1/HIV-2 3rd generation[b]	Abbott Laboratories	S	RP
Enzygnost Anti HIV-1 & 2	Behringwerke	I	SP
Biotest Anti-HIV-1/2	Biotest Diagnostics	I	RP
HIV-1/HIV-2 Modul-Test	Biochrom	I	SP
Peptide HIV ELISA	Cal-Tech Diagnostics	I	SP
HIV (1+2) Ab EIA	Clonatec	I	SP
ELAVIA mixT	Diagnostics Pasteur/Sanofi	I	L
Rapid ELAVIA mixT	Diagnostics Pasteur/Sanofi	I	L

APPENDIX A (Continued)
Retroviral Tests[a]

Name of test	Manufacturer (alphabetical)	Type of Test	Antigen
HIV ELISAs			
For the Detection of Antibody to HIV-1 and HIV-2 (Cont.)			
Genelavia mixture	Diagnostics Pasteur/Sanofi	I	SP/RP
DuPont HIV-1/HIV-2 ELISA	DuPont de Nemours	I	SP/RP
HIV-1/HIV-2 EIA[b]	Genetic Systems/Sanofi	I	L
Anti-HIV-1/HIV-2 EIA	Hoffmann-LaRoche	I	RP
Human HIV 1 & 2	Human	I	RP
Detect-HIV-1/2	IAF Biochem/Coulter	I	SP
Select HIV1/2 Diff.	IAF Biochem/Coulter	I	SP
Innotest HIV-1/2	Innogenetics	I	SP/RP
Vironostika HIV-1 + 2	Organon Teknika	I	L/SP
Wellcozyme HIV-1 + 2	Wellcome Diagnostics	S	SP/RP
GACELISA	Wellcome Diagnostics	S	RP
For the Detection of Antibodies to HIV-1, HIV-2, HTLV-I, and HTLV-II			
Bioelisa HIV-1+ 2, HTLV-I + II	Biokit Ltd.	I	SP
Detect Plus	IAF Biochem International	I	SP
HIV Rapid and Simple Assays			
For the Detection of Antibody to HIV-1			
Retro Cell	Abbott Laboratories	A	L
SimpliRed HIV-1 Ab	Agen	A	SP
Recombigen HIV-LA[b]	Cambridge Biotech	A	RP
HIVCHEK	DuPont de Nemours	Dot	RP
Serodia HIV	Fujirebio	A	L
SUDS	MUREX	Dot	L/SP
PATH HIV Dipstick	PATH	Dot	SP
Immunocomb	PBS Organics	Dot	SP
Serion Immuno Tab HIV-1	Serion Immunodiagnostics	Dot	L
GACPAT	PHLS Laboratories	A	L
For the Detection of Antibody to HIV-1 and HIV-2			
Test Pack HIV-1/HIV-2 AB	Abbott Laboratories	Dot	RP
Recombigen Rap. Test Dev.	Cambridge Biotech	Dot	RP
Rapid HIV-1/2	Clonatec	Dot	SP
HIV-SPOT	Diagnostic Biotechnology	Dot	SP/RP
HIVCHEK 1 + 2	DuPont de Nemours	Dot	SP/RP
Genie HIV-1/2	Genetic Systems/Sanofi	Dot	SP
SUDS 1 + 2	MUREX	Dot	L/SP
Immunocomb Bi-Spot	PBS Organics	Dot	SP

APPENDIX A (Continued)
Retroviral Tests[a]

Name of test	Manufacturer (alphabetical)	Type of Test	Antigen
Recodot	Waldheim Pharmazeutika	Dot	RP
HIV Supplemental Assays			
Envacore	Abbott Laboratories	C	RP
Matrix HIV-1/2	Abbott Laboratories	Dot	RP
Ancoscreen	Ancos	Blot	L
NovaPath[b]	BioRad	Blot	L
Speedscreen HIV-1/2	British Biotechnology	LIA	RP
HIV-1 Western blot[b]	Cambridge Biotech	Blot	L
HIV-2 Western blot	Cambridge Biotech	Blot	L
HIV-2 IFA	Cambridge Biotech	IFA	IC
Retro-tek IFA/HIV	Cellular Products	IFA	IC
RIBA HIV	Chiron	Blot	RP
PEPTI-LAV 1-2	Diagnostics Pasteur/Sanofi	LIA	SP
New LAV blot I	Diagnostics Pasteur/Sanofi	Blot	L
New LAV blot II	Diagnostics Pasteur/Sanofi	Blot	L
HIV-1 Western blot 1.1	Diagnostic Biotechnology	Blot	L
HIV-2 Western blot 1.2	Diagnostic Biotechnology	Blot	L
HIV-1/2 blot 2.2	Diagnostic Biotechnology	Blot	L/SP
IgM Western blot	Diagnostic Biotechnology	Blot	L
HIV-1 Western blot[b]	DuPont deNemours/Ortho	Blot	L
Epiblot[b]	Epitope	Blot	L
Genetic Systems (HIV-1)	Genetic Systems/Sanofi	Blot	L
Genetic Systems (HIV-2)	Genetic Systems/Sanofi	Blot	L
INNO-LIA HIV-1/HIV-2	Innogenetics/MUREX	LIA	SP/RP
HIV Western blot[b]	Organon Teknika	Blot	L
Liatek HIV-1/2	Organon Teknika	LIA	SP/RP
Virgo IFA	Pharmacia	IFA	IC
Serofluor IFA	Virion	IFA	IC
Fluorognost HIV-1 IFA[b]	Waldeheim Pharm/Thermascan	IFA	IC
HIV p24 Antibody Assays			
p24 EIA	Abbott Laboratories	I	L
HIV-1 p24 Ab	Coulter	C	L
ELAVIA Anti p24/25	Diagnostics Pasteur/Sanofi	S	L
Vironostika HIV core	Organon Teknika	S	L
Anti-p24	Wellcome Diagnostics	S	L
HIV Urine Antibody Assays			
Urine ELISA	Calypte Biomedical	I	RP
HIV Antigen Assays			
p24 Antigen EIA[b]	Abbott Laboratories	S	

APPENDIX A (Continued)
Retroviral Tests[a]

Name of test	Manufacturer (alphabetical)	Type of Test	Antigen
HIV Antigen Assays			
p24 EIA	Cellular Products	S	
HIV-1 p24 Antigen EIA	Coulter	S	
ELAVIA Antigen 1	Diagnostics Pasteur/Sanofi	S	
p24 Antigen	DuPont de Nemours	S	
p24 Antigen	Genetic Systems/Sanofi	S	
Vironostika p24 Antigen	Organon Teknika	S	
p24 Antigen	Wellcome Diagnostics	S	
Assays to Detect Antibodies to HTLV-I and HTLV-II			
Screening Assays			
HTLV-I EIA[b]	Abbott Laboratories	I	L
Recombinant HTLV-I EIA[b]	Cambridge Biotech	I	RP
Retrotek HTLV-I ELISA[b]	Cellular Products	I	L
Detect HTLV-I/II	Coulter/IAF Biochem	I	SP
Select HTLV-I/II	Coulter/IAF Biochem	I	SP
HTLV EIA	Diagnostic Biotechnology	I	L
HTLV-I ELISA[b]	DuPont/Ortho	I	L
Serodia HTLV-I	Fujirebio	A	L
Vironostika HTLV-I EIA	Organon Teknika	I	L
Synth EIA HTLV	UBI-Olympus	I	SP
Supplemental Assays for HTLV-I and HTLV-II			
HTLV-I/II Western blot	Cambridge Biotech	Blot	L/RP
Retro-Tek HTLV-I	Cellular Products	Blot	L
Retro-Tek HTLV-I IFA	Cellular Products	IFA	IC
HTLV-I/II Antigen Assay	Coulter/IFA Biochem	S	
HTLV blot (2.2)	Diagnostic Biotechnology	Blot	L/RP
HTLV-I Western blot	DuPont/Ortho	Blot	L
Problot	Fujirebio	Blot	L
HTLV Western blot	Organon Tecknika	Blot	L
Sero-Fluor IFA-HTLV-I	Virion	IFA	IC
Sero-Fluor IFA-HTLV-II	Virion	IFA	IC

Key: C = competitive ELISA, L = lysate, RP = recombinant protein, SP = synthetic peptide, IC = infected cells, I = indirect ELISA, S = sandwich ELISA, A = agglutination assay, Dot = dot-blot or microparticle assay, Blot= Western blot, LIA = line immunoassay, F = fluorometric, CAP = capture assay, and IFA = indirect fluorescent assay.

[a] May not include all available tests.
[b] Licensed by the FDA.

APPENDIX B
Major Gene Products of the Human Retroviruses

Gene	HIV-1	HIV-2	HTLV-I
env			
Precursor	gp160	gp140	gp61/68
External	gp120	gp105 (125)	gp46
Transmembrane	gp41	gp36 (41)	
gag			
Precursor	p55	p56	p53
Core	p24	p26	p24
Matrix	p17	p16	p19
	p9		
	p7		
pol			
RT	p66	p68	–
RT	p51	p53	–
Endonuclease	p31	p34	–

APPENDIX C

OMB Form No.: 0920 0274
Expiration Date: 9/30/92

Model Performance Evaluation Program Enrollment Form

**Model Performance Evaluation Program
Human Immunodeficiency Virus Type 1 (HIV-1),
Human T-Lymphotropic Virus Types I and II (HTLV-I/II),
and T-Lymphocyte Immunophenotyping (TLI)
Laboratory Enrollment Form**

1. Name of Laboratory: _____

2. Mailing Address : _____

3. City: _____ State: _____ ZIP Code: _____

   [ ] Please check if this address differs from that on the mailing label.

4. Telephone Number: (_____) - _____ - _____ Ext: _____

5. Laboratory Director's Name and Title: _____

6. Laboratory Supervisor's Name and Title: _____

7. Please indicate whether your laboratory would like to participate in the CDC Model Performance Evaluation Program for HIV-1, HTLV-I/II, and/or TLI.

HIV-1	HTLV-I/II	TLI
[ ] Yes	[ ] Yes	[ ] Yes
[ ] No	[ ] No	[ ] No

   If Yes, please use the following to describe the **primary classification** of your laboratory (check only **one**):

   [ ] Hospital, non-blood bank
   [ ] Health department
   [ ] Hospital blood bank
   [ ] Non-hospital blood bank
   [ ] Independent laboratory
   [ ] Federal Government research laboratory
   [ ] Drug screening/toxicology laboratory
   [ ] Organ procurement laboratory

   [ ] Employee health clinic
   [ ] University-associated research laboratory
   [ ] Sexually transmitted diseases (STD) clinic
   [ ] Privately funded research laboratory
   [ ] Physician-office laboratory
   [ ] Pharmaceutical laboratory

   [ ] Health Maintenance Organization (HMO)

   [ ] Commercial manufacturer of reagents

   [ ] Other (please specify):_____

   If No, please check the appropriate box below:

       [ ] Our laboratory does not perform HIV-1 testing.
       [ ] Our laboratory does not perform HTLV-I/II testing.
       [ ] Our laboratory does not perform T-lymphocyte immunophenotyping (TLI).
       [ ] Other reasons, please specify (optional):

APPENDIX C

OMB Form No.: 0920 0274
Expiration Date: 9/30/92

Model Performance Evaluation Program Enrollment Form

**Model Performance Evaluation Program
Human Immunodeficiency Virus Type 1 (HIV-1),
Human T-Lymphotropic Virus Types I and II (HTLV-I/II),
and T-Lymphocyte Immunophenotyping (TLI)
Laboratory Enrollment Form - Continued**

8. Please indicate whether your laboratory would like to participate in the Laboratory Performance Information Exchange System (LPIES), an electronic bulletin board system which is available free of charge:

   [ ] Yes
   [ ] No

9. Please send me a copy of the following:

   [ ] Current summary of the HIV-1 antibody testing results.
   [ ] Current summary of the HIV-1 antibody testing laboratory population and their testing practices.
   [ ] Current summary of HTLV-I/II antibody testing results.
   [ ] LPIES Quick Reference Guide

10. Please read the following and sign the form.

    We understand that as participants in the program, we will be asked to send the following to CDC: (1) results of our testing of samples provided by CDC; (2) information on methods used with the samples; and (3) demographic and other information related directly or indirectly to the testing process.

    Director's Signature: _____

11. Please mail the completed enrollment form to:

    MPEP Survey Coordinator
    Program Resources, Inc.
    Science and Technology Center
    P. O. Box 12794
    Research Triangle Park, NC 27709

If you have questions about this program, please contact our CDC office at (404) 639-2137 or write directly to:

Model Performance Evaluation Program
Division of Laboratory Systems
Public Health Practice Program Office
Centers for Disease Control
Building 6, Room 315 MS G23
1600 Clifton Road, NE
Atlanta, GA 30333

If you have any questions about this program or if you know of other facilities of your organization doing HIV-1, HTLV-I/II, OR TLI testing, please call toll free 1-800-322-4383.

APPENDIX D

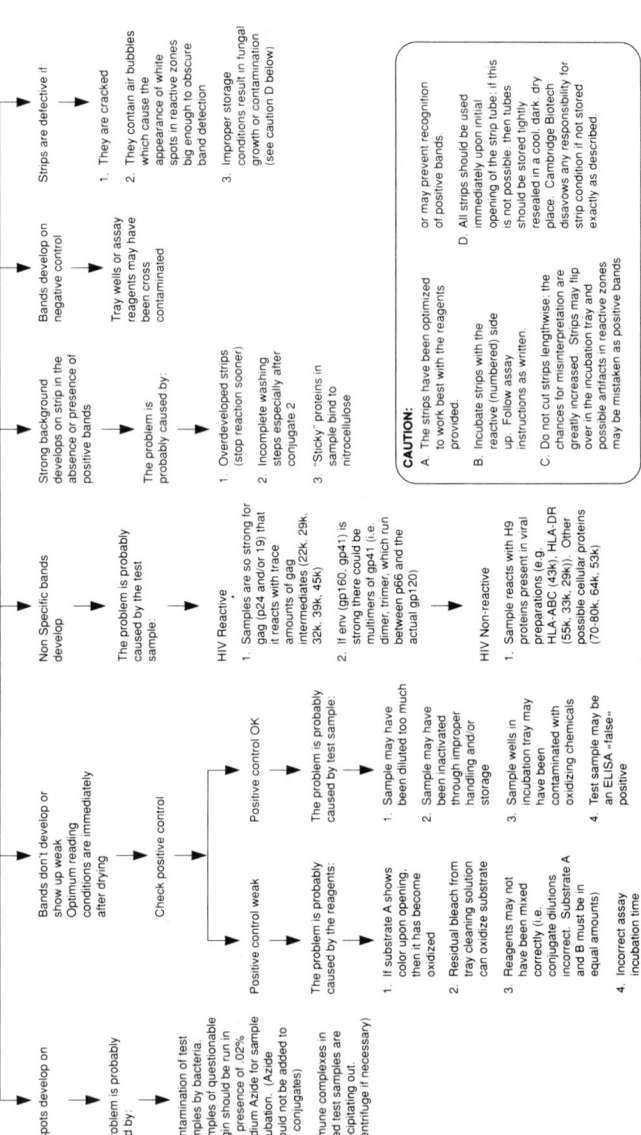

INDEX

A

ADCC, see Antibody-dependent cellular cytotoxicity
Adult lymphoma/leukemia, 103, 104
Agglutination tests, 35, 42–45, 173, 198
AIDS, 16, 23, 28, 69, 198
  in Africans, 69
  antibodies and, 15
  cellular abnormalities in, 30–31
  continuing education on, 132
  functional defects in cells and, 31–32
  HIV antibodies in, 35
  HIV-1 virus and, 7–10
  immune deficiency in, 23
  in infants, 20
  laboratory safety and, 191
  reactivity to p24 antigen in, 80
  survival in, 30
  symptoms of, 9
  wasting syndrome in, 25
AIDS-related complex (ARC), 9, 23
Amino acids, 10, see also specific types
Anemia, 25
Antibodies, 64, see also specific types
  in adults, 15–18
  cross-reacting, 18
  detection of, 35, 37
  to gag protein, 17–18
  to HIV-1, 15–20, 97, 197
  to HIV-2, 90, 91, 97, 197
  to HTLV-I, 197
  to HTLV-II, 197
  immunoglobulin A, 20
  immunoglobulin G, 19, 96
  immunoglobulin M, 19, 96
  in vitro synthesis of, 70–71
  maternal, 19
  monoclonal, 70
  neutralizing, 15
  in newborns, 19–20
  production of, 15
  profiles of, 16, 18, 19
  titers of, 15, 16
  in urine, 51
Antibody capture ELISA, 41–42, 198
Antibody-dependent cellular cytotoxicity (ADCC), 15
Antibody sandwich ELISA, 70, 137, 198
Antigen capture ELISA, 41–42
Antigenemia, 10, 16, 20, 32, 68

Antigen neutralization assays, 70
Antigens, 15, see also specific types
  circulating, 68–70
  env, 92
  of HIV-1, 10–13
  of HIV-2, 90–91
  of HTLV-I, 104–105
  of HTLV-II, 105
  lysate, 35, 105, 107
  recombinant, 36, 48–50, 100
  separation of, 60
  synthetic peptide, 36, 48–50, 100
  transmembrane, 11, 90
Antigen sandwich ELISA, 40–41, 198
Anti-HIV IgA-specific assays, 20
Anti-human immunoglobulin, 37, 38, 61
ARC, see AIDS-related complex
Arthritis, 53
Assays, see also Screening tests; specific types
  agglutination, 35, 42–45, 173, 198
  antigen neutralization, 70
  anti-HIV IgA-specific, 20
  class-specific antibody, 197
  combination, 37, 64, 94, 197
  competitive, 38, see also specific types
  conditions for, 189
  confirmatory, see Supplemental tests
  cytokine, 25–26
  discrepant results between, 189–190
  dot-blot, 46–47, 97
  ENZO, 73
  enzyme-linked immunosorbent, see Enzyme-linked immunosorbent assays (ELISA)
  format of, 36, 40
  functional, 26
  immuno-, see Immunoassays
  immunoglobulin M, 80
  indirect, 38, see also specific types
  indirect fluorescent antibody, see Indirect fluorescent antibody assays (IFA)
  nucleic acid detection, 73–74
  passive hemagglutination, 42
  radioimmunoprecipitation, 59, 67–68, 94
  reference, 187–188
  screening, see Screening tests
  supplemental, see Supplemental tests
  troubleshooting problems in, 161–164
  Western blot, see Western blot assays
ATLL, see Adult lymphoma/leukemia

B

Autoclaving, 193
Autoimmune diseases, 18, 53, 66, see also specific types

B

Bast transformation, 31
B-cells, 21, 26–28, 30, 31
Beta-2 microglobulin, 32
Biohazards, 191, 193–194
Biosafety, 191, see also Safety
Biotin, 61
Biotin-avidin system, 61
Biotinylated anti-human immunoglobulin, 61
Blind testing, 156–158
Blood centrifugation, 172
Blood clotting, 172
Blood collection, 170–171
Blood transfusions, 121, 126
Borderline reactors, 74, 136, 137

C

Cachexia, 25
Capsids, 11
Carbohydrates, 10, see also specific types
CD2, 29
CD3, 27–29
CD4, 7–10, 20, 21, 28
  activation of, 25
  cell-mediated immunity and, 24
  decrease in, 22, 23, 31
  detection of, 27
  HTLV and, 103
  measurement of, 29, 30, 32
CD8, 20–23, 27–29
  decrease in, 31
  measurement of, 30
CD14, 29
CD16, 29
CD19, 29
CD20, 29
CD25, 29
CD38, 29, 30
CD45, 29
CD56, 29
CD57, 29
Cell-mediated immunity (CMI), 23–25, 32
Cellular abnormalities, 25–31
Cellular interactions, 20–23
Centrifugation, 172
Chemotactic factors, 23
Chemotaxis, 23

Children, 19–20, 69
Chromogen, 38
Circulating antigen, 68–70
Class-specific antibody capture ELISA, 42
Clerical errors, 126
CMI, see Cell-mediated immunity
Coefficient of variation, 140–142
Colloidal gold dye, 47
Colony stimulating factors, 23–25
Combination assays, 37, 64, 94, 197, see also specific types
Commercially available tests, 54–56, 84–85, 93–94
Competitive assays, 38, see also specific types
Competitive ELISA, 39–40, 59, 82, 83, 137, 198
Confidentiality, 122
Confirmatory tests, see Supplemental tests
Conjugates, 37, 38, 61
Conserved regions, 11
Continuing education, 132
Cross-reactions, 18, 105
CSF, see Colony stimulating factors
Cultures, 71–72, 80
Cytokine assays, 25–26
Cytokine receptors, 25
Cytokines, 21, 23, 25, 26
Cytoplasmic fluorescence, 77
Cytotoxic T-cells, 21

D

Diagnostic tests, see Screening tests
Dilution errors, 160
Dilutions, 169, 179–182, 184–186
DNA, 5, 8, 10, 11, 73
Dot-blot assays, 46–47, 97

E

EIA, see Enzyme immunoassays
ELISA, see Enzyme-linked immunosorbent assays
Env antigens, 92
Env gene, 90, 104, 105, 107, 110
Env glycoproteins, 10–12
ENZO assay, 73
Enzyme immunoassays (EIA), 38
Enzyme-linked immunosorbent assays (ELISA), 32, 35, 50, 51, 197, see also specific types
  algorithms for, 82, 83
  antibody capture, 41–42

Index

antibody sandwich, 70, 137, 198
antigen capture, 41–42, 198
antigen sandwich, 40–41, 198
combining other assays with, 84
competitive, 39–40, 59, 82, 83, 137, 198
format of, 40
in HIV-2 detection, 92, 95
in HTLV detection, 105
indirect, see Indirect ELISA
interpretation of results of, 75
noncompetitive, see Indirect ELISA
principles of, 37–42
quality assurance and, 129, 134
quality control and, 136
quantitative, 25
sandwich, 40–41, 70, 137, 198
sensitivity of, 38, 132
specificity of, 132
specimen processing for, 173
troubleshootiing problems in, 164
types of, 198, see also specific types
Enzymes, 38, 90, 185, see also specific types
Equipment maintenance and calibration, 158–160
Errors, 160–161, see also specific types
clerical, 126
dilution, 160
labeling, 125
system, 132
technical, 133
transcriptional, 121, 126
External control, 135–137, 141, 142, 147, 149, 154
acceptable control values and, 137, 138
problems in, 165–166

F

False negatives, 17, 121, 133
False positives, 19, 48, 49, 52, 59, 82
antibodies and, 113
in IFA, 66
quality and, 121, 133
reference tests and, 187
Fc receptor, 32
Filter paper, 171
FITC, see Fluorescein isothiocyanate
Flow cytometry, 26, 27
Fluorescein isothiocyanate (FITC), 27
Fluorescence, 65, 77
Fluorochromes, 27, 38, 66
Fluorometric microparticle principle, 47
Fluorometry, 47

Follow-up testing, 127–129
Functional assays, 26
Functional defects in cells, 31–32

G

Gag gene, 89, 90, 104, 107, 110
Gag protein, 10–12, 17–18
Gelatin particles, 42
Gene products, 10, 13, see also specific types
Genes, see also specific types
  env, 90, 104, 105, 107, 110
  gag, 89, 90, 104, 107, 110
  pol, 89, 90, 104
  regulatory, 5, 104
  rev, 91
  structural, 5, 90
  tat, 91
  tax, 104, 105, 107, 110
  vpx, 91
Genomes, 8, 13, 73, 104, 107, see also specific types
Glassware, 178–179
Glycoproteins, 10–12, 76, 90, see also specific types
GM, see Granulocyte-macrophage
Granulocyte-macrophage colony stimulating factor (GM-CSF), 24, 25
Gray zone, 52, 155–156
Growth factors, 23

H

Hairy-cell leukemia, 104
HAM, see HTLV associated meylopathy
Hazardous waste, 193–194
HBV, see Hepatitis B virus
Helper T-cells, 7, 23
Hemolysis, 170, 172
Hepatitis B virus, 191
HIV-1 virus, 5–13, 198
  AIDS and, 7–10
  antibodies to, 15–20, 97, 197
  antigens pf, 10–13
  cellular abnormalities and, 25–31
  cellular interactions and, 20–23
  commercially available tests for, 84–85
  components of, 5, 10
  cytokines and, 23–25
  exposure to, 191, 192
  immune cell functional defects in, 31–32
  immune system during, see Immune system

lymphocytes and, 20–23
nomenclature for, 126
replication of, 5
screening tests for, see under Screening tests
supplemental tests for, see under Supplemental tests
T-cells and, 20–23
HIV-2 virus, 89–101, 198
   antibodies to, 90, 91, 97, 197
   antigens of, 90–91
   commercially available tests for, 84–85
   exposure to, 191, 192
   interpretation of results of testing for, 100–101
   screening tests for, 91–94, 95–97
      commercially available, 55–56, 93–94
   supplemental tests for, see under Supplemental tests
HTLV-I, 18, 103–110, 126
   antibodies to, 197
   antigens of, 104–105
   diseases associated with, 103–104
   screening tests for, 105–107
   supplemental tests for, 107–110
HTLV-II, 18, 103–110, 127
   antibodies to, 197
   antigens of, 105
   diseases associated with, 104
   screening tests for, 105–107
   supplemental tests for, 107–110
HTLV-III, 126
HTLV associated myelopathy (HAM), 103, 104
Human T lymphotropic virus, see HTLV
Hybridization, 73
Hygiene, 191, 194
Hypercalcemia, 104
Hypergammaglobulinemia, 20
Hypervariable regions, 11

I

ICD, see Immune complex dissociation
IFA, see Indirect fluorescent antibody assays
IFN, see Interferons
Ig, see Immunoglobulin
Immune cell functional defects, 31–32
Immune complex dissociation (ICD), 69
Immune deficiency, 16, 23, 31
Immune system, 15–32
   in adults, 15–18
   cellular abnormalities and, 25–31
   cellular interactions and, 20–23
   cytokines and, 23–25
   kinetics of, 15
   lymphocytes and, 20–23
   in newborns, 19–20
   T-cells and, 20–23
Immunity, cell-mediated, 23–25, 32
Immunoassays, see also Assays; specific types
   enzyme (EIA), 38
   line (LIA), 64–65, 84, 100
Immunodeficiency, 16, 23, 31
Immunoglobulin, 37, 38, 61, see also specific types
Immunoglobulin A, 20, 37, 50
Immunoglobulin A antibodies, 20
Immunoglobulin G, 20, 37
Immunoglobulin G antibodies, 19, 96
Immunoglobulin G antibody capture ELISA, 42
Immunoglobulin G antibody capture particle-adherence test, 48
Immunoglobulin M, 17, 20, 37
Immunoglobulin M antibodies, 19, 96
Immunoglobulin M assays, 80
Immunophenotyping, 26
Immunosuppression, 10, 23
Indirect assays, 38, see also specific types
Indirect ELISA, 37, 38, 59, 82, 83, 116, 198
   quality control and, 136
Indirect fluorescent antibody assays (IFA), 59, 65–67, 181
   advantages of, 81
   criteria for positivity in, 77
   in HTLV detection, 110
   laboratory worker training and, 113
Infants, 19–20, 69
Infectious materials, 191–192
Inflammatory conditions, 25
Inner core, 5
In situ hybridization method, 73
Integrase, 6
Interferons, 23–25, 32
Interleukin receptors, 31
Interleukins, 23–25
Internal control, 135–137, 165
In vitro antibody synthesis, 70–71

K

Kinetics, 15, 37, 47

L

Labeling errors, 125
Laboratory efficiency, 129–130
Laboratory safety, 131, 191–195

Index

Laboratory staff evaluation, 131–132
Laboratory staff monitoring, 124–125
Laboratory techniques, 169–186, see also specific types
  for dilution preparation, 169, 179–182
  for glassware, 178–179
  for pipetting, 176–178
  for reagent reconstitution, 178–179
  for specimen collection, 169–171
  for specimen processing, 172–173
  for testing of human injectable preparations, 183–186
  for volumetric measurements, 173–179
Laboratory training programs, 131–132
Latency, 8
Latex particles, 42
LAV, 127
Leukemia, 103, 104
Leukocytes, 21, 32
Leukopenia, 25
Levy-Jennings chart, 144
LIA, see Line immunoassays
Line immunoassays (LIA), 64–65, 84, 100
Long terminal repeat (LTR), 5, 25
LTR, see Long terminal repeat
Lupus, 53
Lymphocytes, 20–23
  B-, 21, 26–28, 30, 31
  HIV-infected, 73
  phenotyping of, 28
  subpopulations of, 30
  T-, see T-cells
Lymphokines, 21, 23
Lymphoma/leukemia, 103, 104
Lymphopenia, 22, 25
Lysate antigens, 35, 105, 107

M

Macrophages, 21, 23, 32
Major histocompatibility complex (MHC), 8
Maternal antibody, 19
MATRIX, 64, 99
Mediators, 21
Membrane-bound cytokines, 26
Messenger RNA, 5, 12, 26
MHC, see Major histocompatibility complex
Microbeads, 37, 42
Microglobulin, 32
Microparticles, 47
Monoclonal antibodies, 70
Monocytes, 23, 32
Monokines, 23
mRNA, see Messenger RNA

N

Natural killer cells, 21, 22, 28–30
Nef protein, 12
Negative predictive value (NPV), 117
Neopterin, 32
Neutralizing antibodies, 15
Newborns, 19–20, 69
NF-kB, 25
NHS, see Normal human serum
Nitrocellulose, 61
4-Nitrophenylphosphate, 38
NK, see Natural killer
Noncompetitive ELISA, see Indirect ELISA
Nonspecific reactions, 108
Normal human serum (NHS), 136
Northern blot tests, 26
NPV, see Negative predictive value
Nucleic acids, 10, 73–74, see also specific types
Nucleocapsids, 11

O

Occupational risks, 191–194
OD, see Optical density
Opportunistic infections, 9, 21, 104, see also specific types
Optical density readings, 38
Outer envelope, 5

P

Pancytopenia, 104
Parallel testing, 126
Passive hemagglutination assays, 42
PBMCs, 73
PCR, see Polymerase chain reaction
PE, see Phycoerythrin
Pediatric HIV, 19–20
Performance evaluation, 156
Personal hygiene, 191, 194
PHA, see Phytohemagglutinin
Phenotyping, 26, 28
Phycoerythrin, 27
Phytohemagglutinin (PHA), 31
Pipettes, 173–178
Pipetting techniques, 176–178
Plasma cells, 21
Plasma collection, 169
PMN, see Polymorphonuclear leukocytes
Pokeweed mitogen (PWM), 70
Pol gene, 89, 90, 104
Pol protein, 10–12
Polymerase, 10, 11

Polymerase chain reaction (PCR), 20, 73, 74, 80, 90, 107, 110
Polymorphonuclear leukocytes (PMN), 21, 32
Positive predictive value (PPV), 117
PPV, see Positive predictive value
Primers, 73
Proficiency panels, 122, 156–158
Proficiency testing, see Quality assessment
Proteases, 6, 11, see also specific types
Protein A, 47
Proteins, 10, 90, see also specific types
  gag, 10–12, 17–18
  nef, 12
  pol, 10–12
  recombinant, 64, 107
  regulatory, 5, 12
  rev, 12
  tat, 12
  vif, 12
  vpr, 12
  vpu, 12
Prozone, 44
PWM, see Pokeweed mitogen

Q

QA, see Quality assurance
QC, see Quality control
Quality assessment, 121, 122, 156–158
Quality assurance, 53, 122–134, 156, 197, 198
  defined, 122
  equipment maintenance and calibration and, 158–160
  factors included in, 123
  follow-up testing and, 127–129
  importance of, 121
  laboratory efficiency and, 129–130
  laboratory staff evaluation and, 131–132
  laboratory staff monitoring and, 124–125
  laboratory vigilance and, 125
  necessity of, 121
  parallel testing and, 126
  physicians and, 127
  record keeping and, 123–124
  results reporting and, 126–127
  scope of, 169
  specimen storage and, 127–129
  supervisors of, 130
  total quality management and, 130–134
  transcriptional measure reviews and, 126
  troubleshooting and, 161–166
  true positive and negative verification and, 125–126
Quality control, 125, 130, 134–156, 197
  acceptable control values in, 137–138, 153–155
  calculated values in, 145–148
  calculations in, 138–142
  coefficient of variation and, 140–142
  defined, 122, 134–135
  duplicate controls in, 152–153
  equipment maintenance and calibration and, 158–160
  external, 135–137, 141, 142, 147, 149, 154
    acceptable control values and, 137, 138
    problems in, 165–166
  graphs for, 143–145
  gray zone reactors and, 155–156
  importance of, 121
  internal, 135–137, 165
  necessity of, 121
  policy in, 151
  protocol for, 151–152
  recordkeeping and, 135
  shifts in, 148–151
  standard deviation and, 139–140, 142–144, 152
  supervisors of, 130
  trends in, 148–151
  troubleshooting and, 161–166
Quantitative ELISA, 25

R

Radioimmunoassays (RIA), 32
Radioimmunoprecipitation assays (RIPA), 59, 67–68, 94, 107, 108
RBCs, see Red blood cells
Reagent reconstitution, 178–179
Receptors, 7, see also specific types
Recombinant antigens, 36, 48–50, 100
Recombinant proteins, 64, 107
Record keeping, 123–124, 135
Red blood cells (RBCs), 42, 45
Reference tests, 187–188
Regulatory genes, 5, 104
Regulatory proteins, 5, 12
Reverse transcriptase (RT), 5, 6, 72
Reverse transcription, 11
Rev gene, 91
Rev protein, 12
RF, see Rheumatoid factor
Rheumatoid arthritis, 53
Rheumatoid factor (RF), 19–20
RIA, see Radioimmunoassays
RIPA, see Radioimmunoprecipitation assays
RNA, 5, 8, 10–12, 26, 73
RT, see Reverse transcriptase

Index

S

Safety, 131, 191–195
Saliva tests, 50–51
Sandwich ELISA, 40–41, 70, 137, 198
Screening tests, 35–56, see also Assays; specific types
  commercially available, 54–56, 93–94
  delta values for, 115–116
  efficiency of, 115
  evaluation of kits for, 187–190
  for HIV-1
    agglutination, 35, 42–45
    commercially available, 54–56
    dot-blot, 46–47
    first-generation, 36
    HIV-2 tests combined with, 95–97
    interpretation of results of, 51–53
    saliva, 50–51
    second-generation, 36
    selection of, 53–54
    third-generation, 36
    urine, 50–51
  for HIV-2, 91–97
    commercially available, 55–56, 93–94
  for HTLV, 105–107
  kits for, 187–190
  predictive values for, 117–118
  reactive classification in, 125
  sensitivity of, 113–115
  specificity of, 114–115, 118
  value of, 113–118
SDS-PAGE, see Sodium dodecyl sulfate-polyacrylamide gel electrophoresis
Secretory immunoglobulin A, 50
Sensitivity, 38, 113–115, 132
Serial dilution technique, 180–182
Seroconversion, 16, 18, 79, 155
Seroprogression, 79
Serum collection, 169
Severe immunosuppression, 10, 23
Shewhart chart, 144
Skin lesions, 104
Sodium dodecyl sulfate-polyacrylamide gel electrophoresis (SDS-PAGE), 60
Specificity, 59, 114–115, 118, 132
Specimens, see also specific types
  choices of, 188–189
  collection of, 169–171
  processing of, 172–173
  risks associated with handling of, 191–194
  storage of, 127–129
Spectrophotometry, 38
Spikes, 5
Spills, 194

Standard deviation, 139–140, 142–144, 152
Structural genes, 5, 90
Substrates, 38, 197, see also specific types
Sugars, 10, see also specific types
Supplemental tests, 59–85, see also specific types
  commercially available, 84–85
  for HIV-1
    advantages of, 80–82
    algorithms for, 82–84
    circulating antigen detection and, 68–70
    commercially available, 84–85
    criteria for positivity in, 75–80
    cultures and, 71–72, 80
    disadvantages of, 80–82
    HIV-2 tests combined with, 97–100
    inconclusive results in, 75, 76, 79–80
    indirect fluorescent antibody assays as, 59, 65–67, 77, 81
    interpretation of results of, 74–80
    line immunoassays as, 64–65, 84
    nucleic acid detection assays as, 73–74
    radioimmunoprecipitation assays as, 59, 67–68
    specificity of, 59
    use of, 59
    Western blots as, see Western blot assays
  for HIV-2, 94–95
    commercially available, 84–85
    HIV-1 tests combined with, 97–100
    in vitro antibody synthesis and, 70–71
  for HTLV, 106–110
Suppressor T-cells, 21, 23
Syncytia, 8
Synthetic peptide antigens, 36, 48–50, 100
Synthetic peptides, 64, 107
Syringes, 170–171
System errors, 132

T

Tat gene, 91
Tat protein, 12
Tax gene, 104, 105, 107, 110
T-cells, 9, 10, 20–23, 29, 30, see also specific types
  cytotoxic, 21
  helper, 7, 23
  proliferation of, 32, 104
  quantitation of, 26–28
  suppressor, 21, 23
Technical errors, 133
Titers of antibodies, 15, 16
TNF, see Tumor necrosis factor

Topical spastic paraparesis (TSP), 103–104
Total quality management (TQM), 130–134, 187
TQM, see Total quality management
Transcription, 8, 11
Transcriptional errors, 121, 126
Transcriptional measure reviews, 126
Transfusions, 121, 126
Translation, 8
Transmembrane antigen, 11, 90
Troubleshooting, 161–166
True negatives, 125–126
True positives, 125–126
TSP, see Topical spastic paraparesis
Tumor necrosis factor, 23, 25

U

Urine collection, 169
Urine tests, 50–51

V

Vacuum tube devices, 171
Variable regions, 11, 15
Vif protein, 12
Volumetric measurements, 173–179
Vpr protein, 12
Vpu protein, 12
Vpx gene, 91

W

Wasting syndrome, 25
WB, see Western blot
Western blot assays, 51, 59, 82, 197
    criteria for positivity in, 75–78, 108, 109
    disadvantages of, 81
    in HIV-2 detection, 94, 95, 97, 100
    in HLTV detection, 107–110
    principles of, 60–64
Window period, 17